AF861504

# RADIO BUILDERS' MANUAL

**COMPLETE PLANS FOR**

**ALL-WAVE RECEIVERS** — **AMATEUR TRANSMITTERS**
**POLICE CALL ADAPTERS** — **CRYSTAL SETS**
**AUTOMOBILE RADIO** — **PORTABLE RECEIVERS**

as well as hundreds of other radio plans,
stunts, and trouble-hunting kinks.

reprinted by Lindsay Publications Inc, Bradley IL 60915 • all rights reserved • ISBN 1-55918-116-8

# Radio Builder's Manual

Originally published by
Modern Mechanix Publishing Company
Minneapolis MN

Original copyright 1935 by
Modern Mechanix Publishing Company

Reprinted by
Lindsay Publications Inc
Bradley IL 60915

All rights reserved.

ISBN 1-55918-116-8

1 2 3 4 5 6 7 8 9 0

1993

**WARNING**

Remember that the materials and methods described here are from another era. Workers were less safety conscious then, and some methods may be downright dangerous. Be careful! Use good solid judgement in your work, and think ahead. Lindsay Publications Inc. has not tested these methods and materials and does not endorse them. Our job is merely to pass along to you information from another era. Safety is your responsibility.

Write for a complete catalog of unusual books available from:

Lindsay Publications Inc
PO Box 12
Bradley IL 60915-0012

# I'LL TRAIN YOU AT HOME *In Your Spare Time For A* GOOD RADIO JOB

J. E. SMITH, President
National Radio Institute

**The man who has directed the Home-Study Training of more men for the Radio industry than anyother man in America.**

Mail the coupon now. Get the facts about Radio—the field with a future. N. R. I. training fits you for jobs in connection with the manufacture, sale and operation of Radio equipment. It fits you to go in business for yourself, service sets, operate on board ships, in broadcasting, television, aviation, police Radio and many other jobs. My FREE book tells how you quickly learn at home in spare time to be a Radio Expert.

### Many Radio Experts Make $40, $60, $75 a Week

Why struggle along in a dull job with low pay and no future? Start training now for the live-wire Radio field. I have doubled and tripled salaries. Hundreds of successful men now in Radio got their start through N. R. I. training.

### Many Make $5, $10, $15 a Week Extra in Spare Time While Learning

Hold your job. I'll not only train you in a few hours of your spare time a week, but the day you enroll I start sending you Extra Money Job Sheets which quickly show you how to do Radio repair jobs common in most every neighborhood. I give you Radio Equipment for conducting experiments and making tests that teach you to build and service practically every type of receiving set made. Cleo T. Retter, 30 W. Beechwood Ave., Dayton, Ohio, wrote: "Working only in spare time, I made about $1,500 while taking the Course."

### Find Out What Radio Offers

My book has shown hundreds of fellows how to make more money and win success. It's FREE to any ambitious fellow over 15 years of age. Investigate. Find out what Radio offers you. Read what my Employment Department does to help you get into Radio after graduation, about my Money Back Agreement, and the many other N. R. I. features. Mail the coupon in an envelope, or paste it on a 1c post card TODAY.

J. E. SMITH, President, Dept. 5CO1
National Radio Institute, Washington, D. C.

## HERE'S PROOF that my training pays

**Now Owns Own Business**
"If I had not taken your Course I would be digging ditches instead of running my own business. One week I made $75 on repairing alone and this doesn't count sales. If a fellow wants to get into Radio, N. R. I. is the starting point." **R. S. Lewis,** Modern Radio Service, Pittsfield, Ill.

**Spare Time Jobs Earn $15 a Week**
"I have no trouble getting Radio work. I have the reputation of being the best Radio man in town, and average $15 to $20 a week for spare time only." **G. Bernard Croy,** 151 Washington St., Bronson, Michigan.

**Owes His Success to Radio Training**
"Since 1929 I have earned my living in Radio. I owe my last three jobs to N. R. I. I am now in the main control room of one of the large broadcasting chains." **Serge A. De Somov,** 1516 Library Ave., New York City.

**MAIL THIS NOW**

J. E. SMITH, President, Dept. 5CO1
National Radio Institute, Washington, D. C.

Dear Mr. Smith: Without obligating me, send your book which points out the spare time and full time job opportunities in Radio and your 50-50 method of training men at home in spare time to become Radio Experts.

*(Please print plainly)*

NAME .................................................... AGE........

ADDRESS ....................................................

CITY ........................................ STATE....................

# RADIO BUILDERS' MANUAL

PUBLISHED BY
MODERN MECHANIX PUBLISHING COMPANY
529 SOUTH SEVENTH STREET, MINNEAPOLIS, MINNESOTA
(Copyright 1935 by Modern Mechanix Pub. Co.)

## TABLE OF CONTENTS



*(Continued on page 6)*

*a True*
# SUCCESS STORY
*by an* **R·T·I·** TRAINED MAN

# *Could you use an* EXTRA $200 TOO?

## THEN GET INTO RADIO!

*WHERE MANY MEN I HAVE TRAINED ARE MAKING*

## $35 TO $75 A WEEK

***My big FREE Book tells you how. Mail the coupon now. This book has shown hundreds of fellows how to quickly make more money and win success.***

"It was through R-T-I Training that I got the job I now have as Radio Expert for the Oklahoma City Police and Fire Departments. The Course has been of great benefit to me in the practical work that I am doing, and a raise in salary has made me more enthusiastic than before.

"In addition to this, I have made about $200 doing spare time radio work since I enrolled. At present, although I am very busy with my police work, I am able to make from $5 to $8 a week extra in this way."

J. F. BEGGS,
428 West California Ave.,
Oklahoma City, Okla.

**HE CLIPPED THE R-T-I COUPON WHY DON'T YOU?**

### Shows How I Train You at Home In Your Spare Time for a Good-Pay Radio Job

Why struggle along in a small-pay, no-future job? Get into Radio—the field with a future. I'll train you at home, in your spare time. Keep your present job and income while training. What's more, I'll show you how to earn extra money in your spare time during your Training period.

### Many Make $5 to $20 a Week Extra in Spare Time . . . While Learning

That's what many of my students do soon after enrolling. Delmer Grote, Greenville, Ohio, wrote, "Made my first spare-time money six weeks after enrolling. Earned over $500 on spare-time work in ten months while studying."

### Radio Is Full of Real Opportunities for Trained Men

But only "trained" men can qualify. R-T-I Training fits you for Radio's glorious opportunities—to service sets—to hold a good-pay job in broadcasting, aviation, television, police Radio and many other fields rich in opportunities.

### Learn Radio from Real Radio Engineers

R-T-I Training is practical—proven. Famous Radio engineers, working right in the heart of the Radio industry, direct and supervise it. You get the help and advice of these famous men. You get Radio Equipment these engineers selected and approved for conducting experiments and making tests that teach you the practical side of Radio.

### Five Advanced Courses, Without Extra Cost

I help you specialize for Radio's biggest jobs—and don't charge you a penny extra for the advanced training. I give you extra—FIVE Advanced Courses that are being sold right now for $39.50 each. These take you into the latest developments of Television, Sound Pictures, Photo Cells and Public Address Systems.

### My FREE Book Gives You All the Facts

Tells about Radio's spare-time and full-time opportunities. About my hundreds of successful students. About my employment department. About my Money Back Agreement. About my Four Big Outfits of Radio Equipment furnished without extra cost.

Clip and mail the coupon now. Learn how you can get into a field with a real future. Learn why 45 big radio manufacturers endorse and recommend R-T-I Training . . . how the famous R-T-I Advisory Board works for you and with you while studying and after graduation.

**RAY D. SMITH,**
**President,**
**RADIO & TELEVISION INSTITUTE,**
**Chicago**

**R-T-I'S Famous Advisory Board of Nationally Known Radio Engineers Will Supervise and Direct Your Training**

**Kendall Clough**
Chief Engineer,
Clough-Brengle Company.

**E. E. Gramer**
Chief Engineer,
Standard Transformer Corporation.

**Karl E. Hassel**
Chief Engineer,
Zenith Radio Corporation.

**H. C. Tittle**
Chief Radio Engineer,
General Household Utilities (Grunow).

**F. D. Whitten**
Service Manager,
Chicago Branch,
Philco Radio & Television Corp.

**Dr. C. M. Blackburn**
Radio Engineer,
P. R. Mallory & Co.,
Indianapolis,
Manufacturers of Radio Apparatus.

**MAIL COUPON FOR FREE BOOK**

**RAY D. SMITH, President**
**Radio & Television Institute (R-T-I)**
**2130 Lawrence Ave., Dept. 53**
**Chicago, Ill.**

Dear Mr. Smith: Without obligating me send FREE BOOK about spare-time and full-time Radio opportunities and how I can train for them at home.

Name.............................................................

Address..................................................Age........

City...........................................State............

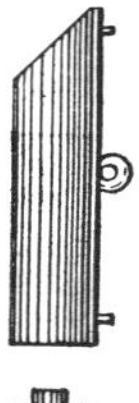

## SPECTROSCOPE

Reveals the secrets of both suns and atoms! Shows the countless spectral lines of the sun and hundreds of brilliant rainbow colored lines of gases and metals. Most amazing instrument analyzes chemicals and ores by their characteristically arranged and beautifully colored spectra. Months of fascinating experiment. New invention makes possible all these achievements at one-tenth the usual cost. Spectroscope complete, **$2.50**

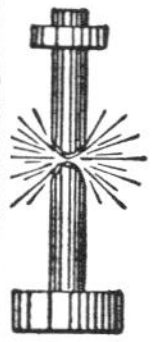

## ELECTRIC FUN

Do tricks with electricity. Make things spin, jump, kick, vibrate, buzz, shoot, shock, flash, mystify—all by electricity. Make window novelties, floating rings, electric lights that obey the voice, spirit rapping, all kinds of amusing and practical devices. Book telling how to do 200 stunts with 110 volts A.C. Price postpaid....**$1**

## STROBOSCOPE

Stroboscope complete with variable speed 110-volt electric motor, neon lamp and book of 100 experiments to be performed with the outfit, telling how to produce strange and beautiful stroboscopic phenomena, motion picture effects, color harmony and many practical applications. Complete outfit, postpaid..................**$1**

**Combination Offer:** With an order for all three of the above items, a copy of **Chemical Fun** (100 experiments) will be included **free.**

**CUTTING & SONS 45 I St. Campbell, Calif.**

## ELECTRICAL SUPPLIES

Complete line—any quantity. Insulating materials; flexible rubber-covered wire; magnet wire; bakelite; fibre rods, sheets, and tubes; insulating paper, tapes, varnishes, sleeving, etc. Send for latest bulletin today.

**AUTO POWER Dept. X 412 S. Hoyne Ave., Chicago**

## For Your Workshop LIBRARY--

HOW TO BUILD IT—1934 Edition—plans for 20-Meter Transmitter, 2-Tube Short Wave Converter, 2-Tube S. W. Receiver, Canoe, Outboard Boat, Water Wheel, Electric Eye, Dirt Track Racer, etc. ....................50c

HOW TO BUILD IT—1932 Edition—plans for Cabin Cruiser, M-M Speedster, Diving Helmet, Ice Yacht, Radios, and scores of other big projects ..............................50c

HANDY MAN'S HOME MANUAL—1st Edition—tells how to repair radios, plumbing, other home equipment, how to build Bungalow Camping Trailer, Electric Refrigerator, Duck-boat, other shop projects..................50c

1931 FLYING AND GLIDER MANUAL—plans for Alco Sportplane, Heath Baby Bullet, Northrop Glider, Airplane Engine, and other light planes and gliders...................50c

1932 FLYING AND GLIDER MANUAL—plans for famous Pietenpol Air Camper Monoplane, Heath Super-Soarer Glider, Powell P. H. Racer, and others........................50c

1933 FLYING MANUAL—plans for T Ford-powered Pietenpol Sky Scout, Gere Biplane, Corben Cabin Monoplane, Cross - Country Twin, Harlequin Motor, etc. ..............50c

HOMES FOR NEWLYWEDS and 16 SELECTED BUNGALOWS FOR LEAN PURSES—two valuable booklets containing illustrations and floor plans of selected low-cost homes—both for......................25c

**All Prices Are Postpaid. Send Your Order To**

**MODERN MECHANIX PUBLISHING CO.**
**529 S. 7th St., Minneapolis, Minn.**


# CONTENTS

*(Continued from page 4)*

## POWER SUPPLIES



## SHORT WAVE RECEIVERS



## STUNTS



## TRANSMITTERS AND MIKES



*(Continued on page 8)*

STUDENTS BUSY AT WORK IN GREAT COYNE SHOPS

# *Learn* RADIO

## I'll Finance Your Training

Yes, sir! I mean what I say. I'll train you in 10 weeks right here in the Great Coyne Shops — NOT BY CORRESPONDENCE — but by actual work on **Radio, Sound** and **Television** equipment AND YOU WON'T HAVE TO START PAYING ME BACK UNTIL TWO MONTHS AFTER YOUR REQUIRED TRAINING PERIOD IS OVER AND THEN YOU'LL HAVE OVER A YEAR TO COMPLETE YOUR PAYMENTS.

# This Is the Greatest Offer Ever Made by Any School

**I MEAN exactly what I say. You can get a complete training by the world's oldest and largest Practical Radio School and I'll finance your Tuition. You won't have to start paying me back in small, monthly payments UNTIL 2 MONTHS AFTER YOUR REQUIRED TRAINING PERIOD IS OVER. I consider the fellow who is ambitious enough to want to get ahead by taking my Training, worthy of my help. MAIL THE COUPON BELOW and you can prove to me that you are willing to spend just 10 weeks in the Coyne Training Shops learning Radio. Then, I'll tell you how I finance your Tuition—give you your complete Training and let you pay me back later.**

# Learn By Doing In 10 Weeks

I don't care whether you are 16 or 45. It makes no difference to me if you don't know an oscillator from a spark coil. You don't need any previous experience or advanced education to master my Shop Training. Don't let lack of money stop you. Most of the students at Coyne have no more money than you have. That's why I have worked out my finance offer.

## MANY EARN While LEARNING

If you need part-time work to help pay your living expenses and will tell me your problems I may be able to help you as I have thousands of others. Then, in 10 brief weeks, in the great Training shops of Coyne, you will learn on a wide variety of modern, up-to-date A.C. Superheterodyne sets, oscillators, analyzers and test instruments. Prepare for Amateur Broadcast, or Telegraph Radio Operator's License and to know all code and Dept. of Commerce rules for a government License Examination.

## TRAINING *By Actual Work*

No dull books, no baffling charts, no classes, you get individual training . . . all real actual work with only the theory you will need. Building real radio sets, doing radio wiring and testing, trouble-shooting, repairing and servicing. That's a glimpse of how I help to make you a Radio expert, and fit you to qualify for jobs leading to the biggest pay.

### Auto Radio Instruction

**Prepare for Jobs Like These**

Here are a few of hundreds of positions in the Radio field. My free employment bureau gives you life-time employment service.

- AIRCRAFT RADIO OPERATOR
- SHIP RADIO OPERATOR
- RADIO SERVICING
- PUBLIC ADDRESS SYSTEMS
- BROADCAST STATION OPERATOR
- TALKING PICTURE STUDIOS
- TELEVISION LABORATORY MAN

## Jobs - Pay - Future

"Two weeks after graduation I received a splendid job. The main consideration given my application was that I was a Coyne Trained man," reports Harold Soucy of Illinois. "I wish to thank your Employment Manager for securing this position for me," writes Albert Yagon, "he sent me out to this Company the first day and I was employed there immediately." And I could go on quoting from hundreds of letters of successful Coyne Trained men. What they have done, you should be able to do!

### Electric Refrigeration Air Conditioning

Right now I'm including my big new Electric Refrigeration and Air Conditioning course at no extra cost.

### *Home of Coyne Shops*

This is our fireproof modern Building wherein is installed thousands of dollars' worth of Radio equipment of all kinds. Every comfort and convenience has been arranged to make you happy and contented during your Training.

## GET THE FACTS

Coyne is your one greatest chance to get into Radio. Every obstacle is removed. This school is 36 years old—Coyne training is tested—proven beyond all doubt—endorsed by many large concerns. You can find out everything absolutely free. Simply mail the coupon and let me send you the big, free Coyne Book with photographs . . . facts . . . jobs . . salaries . . opportunities. Tells you how many earn expenses while training and how we assist our graduates in the field. This does not obligate you. So act at once. Just mail coupon.

**Get This FREE Book**

**H. C. LEWIS, President**
**Coyne Radio & Electrical School**
**500 S. Paulina St., Dept. 35-1M, Chicago**

Dear Mr. Lewis: Without obligation send me your big free catalog and all details; also tell me all about your **"Pay After Graduation Offer."**

*Name*..........................................

*Address*..........................................

*City*.............................*State*..........

H. C. LEWIS, PRESIDENT — FOUNDED 1889
**COYNE RADIO & ELECTRICAL SCHOOL**
**500 S. Paulina St., Dept. 35-1M, Chicago**

Build Your Own Boat!

20 different sets of Plans **50c**

SAILBOATS
CRUISERS . OUTBOARDS
RUNABOUTS . SPEEDBOATS
KYACK . ICEBOAT . ROWBOAT
IN

**How-To-Build 20 Boats**

Order your copy now. Send 50c, stamps or coin, to
MODERN MECHANIX PUBLISHING CO.
529 So. Seventh Street Minneapolis, Minn.

# CONTENTS

*(Continued from page 6)*



*(Continued on page 11)*

# RADIO BUILDERS' HEADQUARTERS

## ALLIED RADIO

is the one dependable source for all of your radio supplies. We specialize in supplying at the lowest prices, matched quality kits for building any type of radio circuit. Send for our valuable current catalog. It is packed with new All-Wave, Dual Wave, and Short Wave kits—thousands of Set-Building parts, tools, test equipment, etc. We have available complete kits for building any **Modern Mechanix** radio device. Send for FREE Parts Lists covering any of the circuits described in the radio pages of **Modern Mechanix** or in this **Radio Builders' Manual.**

## WRITE FOR RADIO'S LEADING CATALOG..

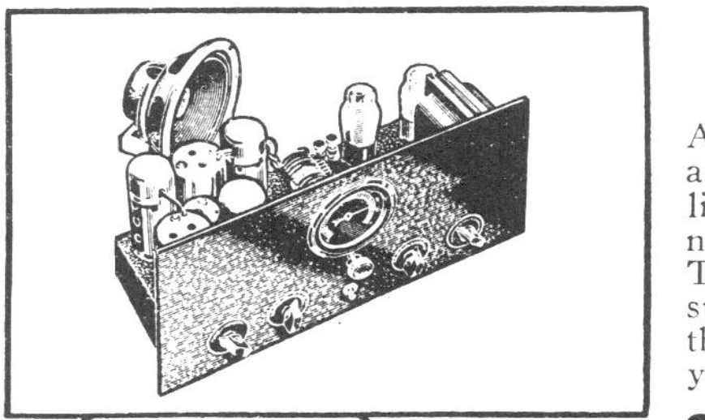

### POPULAR ALL-WAVE KITS
### THE "ALL-STAR" SERIES

The current ALLIED Catalog lists dozens of popular kits. Among these are the sensational ALL-STAR All-Wave Senior and Junior models—described in all of the leading radio publications. Diagrams and Parts Lists for building these famous new radio receivers will be sent to you FREE upon request. Thousands have built the ALL-STAR models with amazing success. You begin with the inexpensive Foundation Kits, and then buy the rest of the parts as you assemble the set, to suit your pocket. It's easy to build the ALL-STAR Kits.

### 2 TUBE SW "DX-ER"

This simple Short Wave Receiver has been built by many radio enthusiasts. It is fully described in the current ALLIED Catalog. Diagrams and Parts Lists are available upon request. The "DX-ER" is very inexpensive to build. Complete instructions are furnished, showing where every wire goes. The radio beginner will be particularly interested in this simple, inexpensive, but efficient Short Wave Receiver.

**DOZENS OF OTHER KITS IN THE CURRENT ALLIED CATALOG**

### 5 METER, 2 WAY AMATEUR PHONE

5 Meter Phone Communication is a fascinating new Amateur Radio activity. Many are building this popular Amateur 2 Way Phone described in the leading Radio publications. Easy to build—very low in cost. Equally successful as transmitter and receiver. Fully described in the current ALLIED Catalog. Write for FREE Parts List. ALLIED is Radio Parts Headquarters for all Short Wave Transmitting and Receiving equipment.

## FREE ALLIED CATALOG

**ALLIED RADIO CORPORATION**
**833 W. Jackson Blvd., CHICAGO, ILL. Dept. B1**

☐ Send me a FREE copy of the current ALLIED Catalog—the leading Radio Supply Guide. I am also interested in diagrams and parts lists for:

☐ ALL-STAR Senior ☐ ALL-STAR Junior
☐ 2 Tube S-W "DX-ER" ☐ 5 Meter Phone

Name ..............................

Address ..............................

# Your Favorite Hobbies

WHETHER you build boats, furniture, radios, models—whether you enjoy stories about aviation, new inventions, outdoor sports, auto racing, or industry, you will find the newest and best articles on your hobby in each issue of

**MODERN MECHANIX AND INVENTIONS**

**15c a copy** **$1.50 a year**

# CONTENTS

*(Continued from page 8)*



# RADIO SET BLUEPRINTS

Standard Modern Mechanix Blueline Prints are available on the following sets described in this book: Hitch-Hiker's 2-Tube Portable—50c; 5-Meter Transceiver—50c; 1-Tube Adapter or Converter—25c; Cakepan Transmitter—25c; 5-Meter Radiophone Transmitter—$1.00; 5-Meter Radiophone Receiver—75c. Send orders to

MODERN MECHANIX PUBLISHING CO.
529 S. 7th St. Minneapolis, Minn.


# RADIO

● holds great rewards for trained men

Almost in a decade radio has become a giant industry. The opportunities created will be enjoyed by trained men. The International Correspondence Schools Radio Course, prepared by leading authorities and constantly revised, will help make you a trained man! A fascinating book—FREE.

## ELECTRICAL ENGINEERING
## —A Great Industry with a Great Future

The new course in Electrical Engineering covers this fascinating field thoroughly. If you are just starting or are partly on the way to a real paying position in this industry, this new course brings the broader knowledge necessary for advancement. The coupon brings complete information.

## MECHANICAL ENGINEERING
## —A Field with a Future

This course has meant advancement to hundreds of ambitious men, mechanics, draftsmen, blacksmiths, foundrymen, patternmakers, apprentices, and mechanical engineers—men who wanted to get ahead and to earn more money. You will find, as they have found, that enrolling for an I. C. S. Course is an investment in a Bigger Future. Use the coupon!

## AUTOMOBILE EXPERTS
## are always in demand

The rapid progress in manufacture and service of the modern automobile brings new opportunities to thousands of ambitious and forward-looking men. There's always a demand for trained men. The new I. C. S. Automobile Course will make you a trained man! The time to act is now. Use the coupon.

INTERNATIONAL CORRESPONDENCE SCHOOLS

BOX 1398, SCRANTON, PENNA.

Without obligation, send me full information on subject checked below:

☐ RADIO ☐ AUTOMOBILE COURSE
☐ ELECTRICAL ENGINEERING
☐ MECHANICAL ENGINEERING

☐ Architecture
☐ Concrete Construction
☐ Reading Shop Blueprints
☐ Civil Engineering
☐ Diesel Engines
☐ Gas Engine Operating
☐ Refrigeration
☐ Chemistry
☐ Highway Engineering
☐ Poultry Farming
☐ Drafting ☐ Accounting
☐ Good English
☐ Salesmanship
☐ Advertising
☐ Show Card Lettering
☐ Civil Service
☐ High School Subjects
☐ Air Conditioning
☐ Inventing and Patenting
☐ Aviation Engines

Name ____________________

Address ____________________

**1935 How To Build It at all newsstands . . . 50c**

# HITCH-HIKER'S

**Weighing only ten pounds complete with batteries, this compact portable receiver with built-in loudspeaker brings in local and distant stations with surprising clarity and volume. Toss the aerial wire into a tree and it's ready for operation.**

**Hitch-hiker's radio is ready for use when insulated aerial is tossed into a tree.**

TO THE camper, the wandering hitch-hiker, the canoeist, or the boating fan this lightweight two-tube receiver, bringing in distant as well as local stations on the built-in loudspeaker and short-wave stations on headphones, will prove a welcome traveling companion.

Where battery box and receiver are connected by a four-wire cable, only one connection, that to an aerial, need be made to get the set in operation. Twenty-five to fifty feet of copper wire, with enough stout cord to pull one end over some nearby tree, will serve admirably as an antenna for this set. The metal chassis, acting as a counterpoise when the set is near the earth, makes a ground connection unnecessary.

A type '32 screen grid detector tube and type '33 power amplifying tube are used in this set, regeneration being controlled by a potentiometer.

The set is built into a 6"x6"x6" electralloy or aluminum shield can, which houses everything but the batteries. It will operate at a cost of about a cent an hour from flashlight batteries. At home, two dry cells

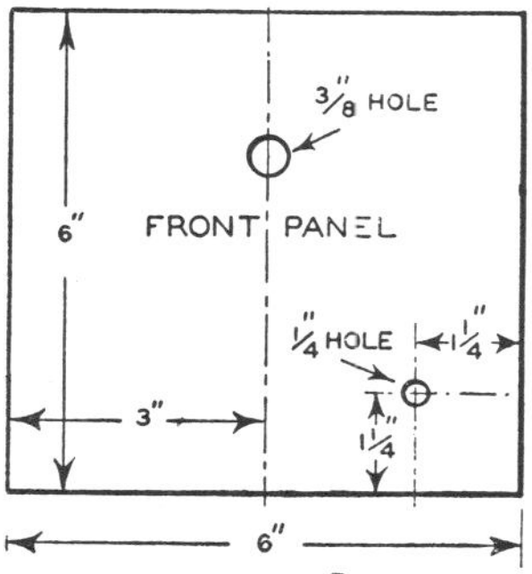


**Front panel layout of set is shown at left. If special vernier tuning dial is used, or airplane type dial, other holes may have to be drilled. If ground wire is used, it may be connected to metal chassis at any point. In many cases, however, better reception is obtained without any ground. Long aerial is needed for short-wave reception—this should be mounted very high.**

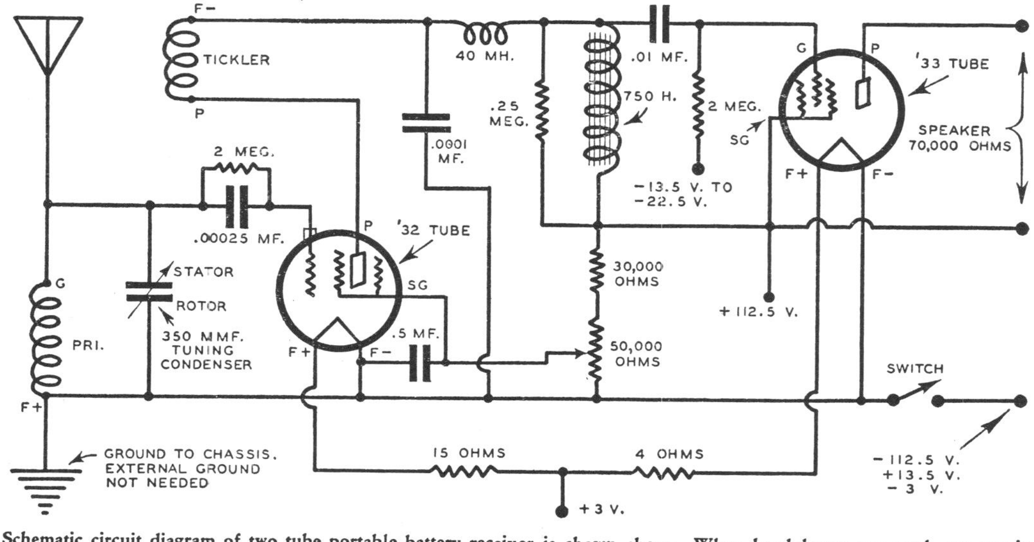


**Schematic circuit diagram of two tube portable battery receiver is shown above. When headphones are used, connect in place of the loudspeaker. Jack may be mounted on front panel if desired, connected so either headphones or speaker can be plugged in without taking cover off set. Mount small resistors and condensers at points convenient for easy wiring.**

# 2-Tube Portable Radio

By
BILL BARTLETT

may be used instead, at a cost of about a cent for three hours.

The aluminum shield can comes knocked down but is easily assembled. Bottom and top screw on, and sides are held in place by channel grooves in the corner posts into which they slide. Solder and bolt a screen door handle or a drawer pull to the top lid for carrying. Drill the front panel to take the dial, carefully following the template shown in the drawing.

The shaft of the potentiometer is insulated from the panel with a fiber bushing. A hole 4½" in diameter is cut in the back side of the chassis to pro-

*(Continued on page 125)*

**PARTS LIST**

Variable cond.—350 mmf.; antenna trimmer cond.; 50,000 ohm midget volume control potentiometer with snap-on filament switch; 750 henry a.f. choke; 60 mh. r.f. choke; 2 four-prong wafer sockets; 1 five-prong wafer socket; vernier tuning dial; volume control knob; one-watt carbon res., 30,000 ohms, .25 megohms, two 2.0 megohms; wire-wound res., 4 and 15 ohms; mica cond., .0001, .00025 mfd.; tubular cond., .01, .5 mfd.; 5" midget magnetic speaker; '32 and '33 tubes; batteries, shield can, plug-in coil form, No. 32 or 36 enameled wire, and hook-up wire.

Coil for broadcast band—200 to 500 meters—is shown below. Wind both primary and tickler in same direction, with turns close together, making the connections to the prongs as indicated.

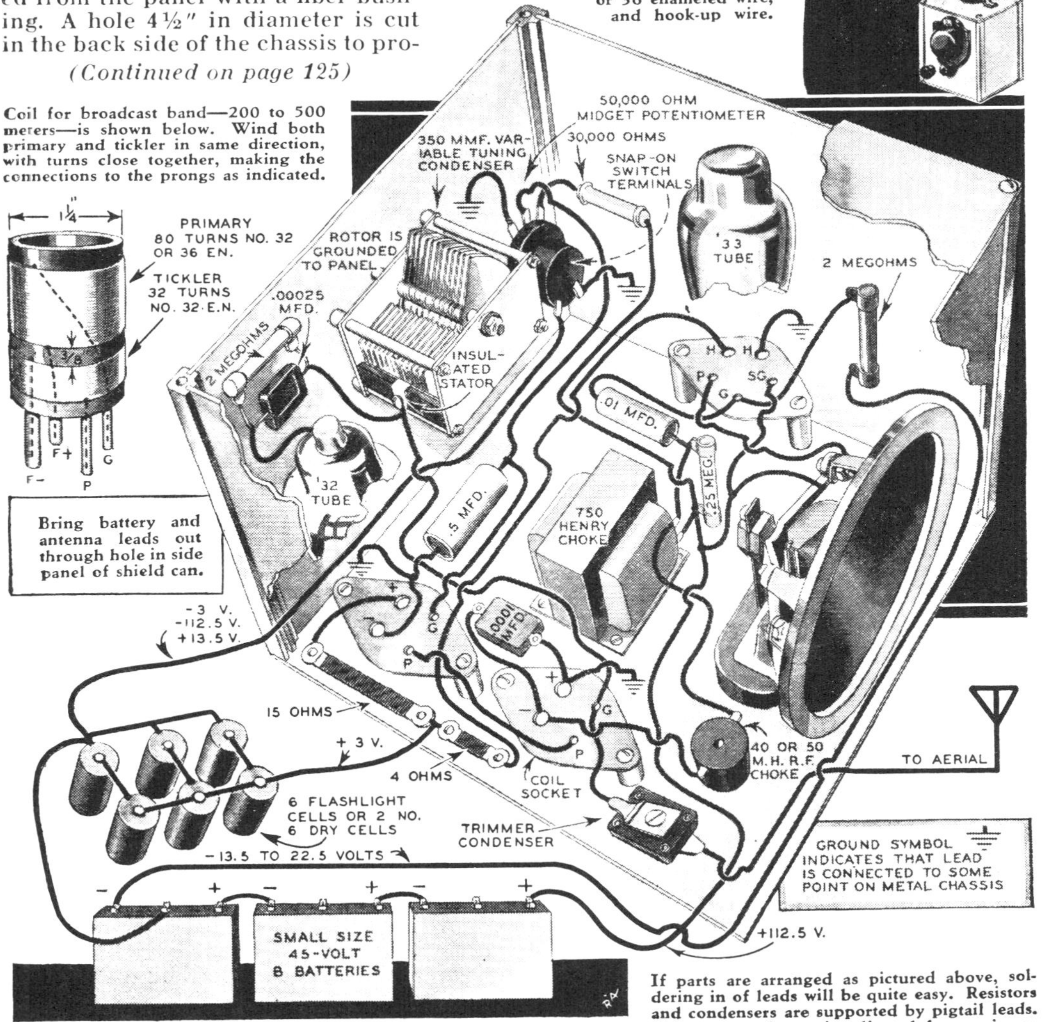


If parts are arranged as pictured above, soldering in of leads will be quite easy. Resistors and condensers are supported by pigtail leads. Trimmer condenser is adjusted for maximum volume only when coils are changed. Three plate vernier condenser may be mounted on front panel in its place.

# Use 5-Meter Transceiver on Car or Boat

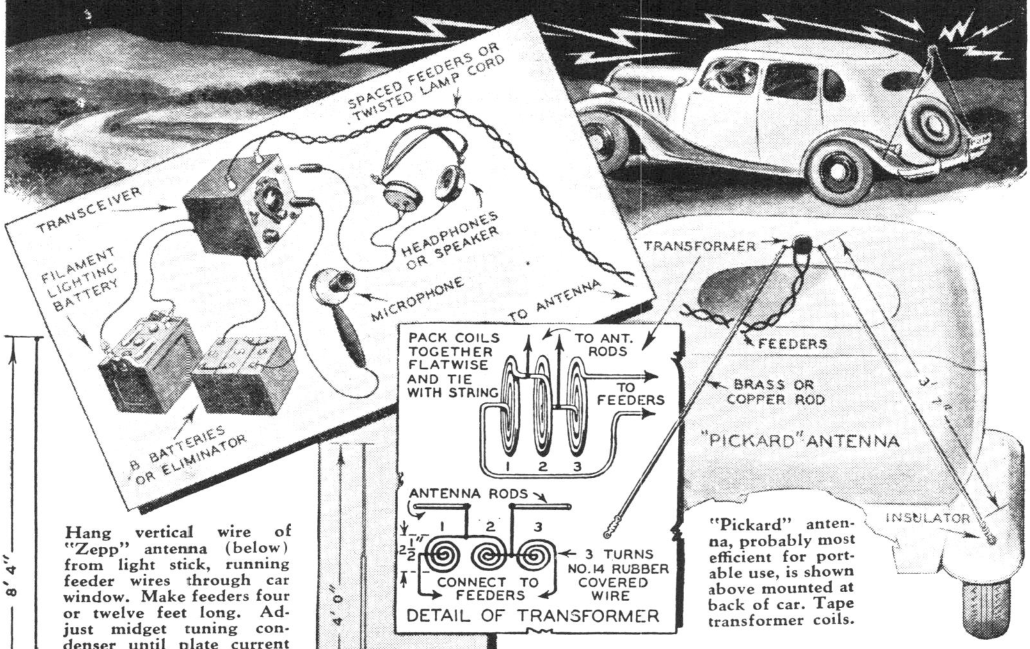


Hang vertical wire of "Zepp" antenna (below) from light stick, running feeder wires through car window. Make feeders four or twelve feet long. Adjust midget tuning condenser until plate current of transceiver shows sharp rise as indicated by meter.

"Pickard" antenna, probably most efficient for portable use, is shown above mounted at back of car. Tape transformer coils.

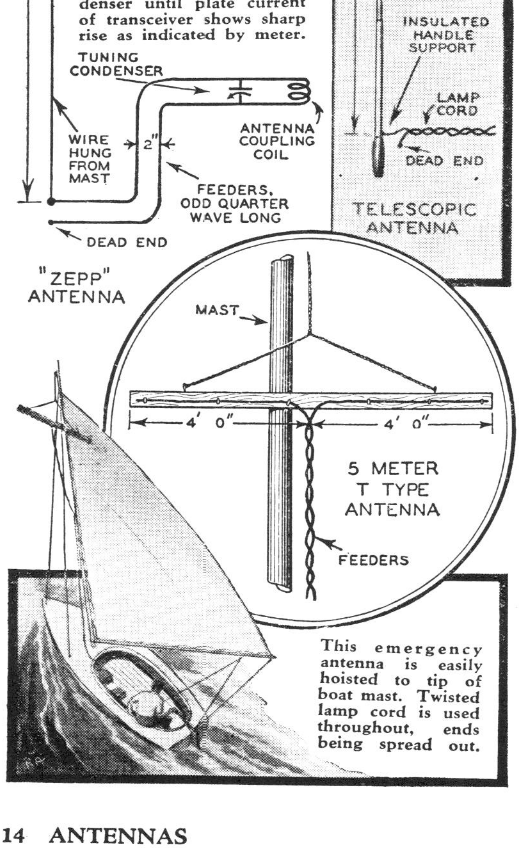


This emergency antenna is easily hoisted to tip of boat mast. Twisted lamp cord is used throughout, ends being spread out.

*by* L. B. ROBBINS

SINCE the advent of the Transceiver unit, 5-meter transmission and reception from both automobiles and boats has become very popular among amateur radio fans. It so happens that the 5-meter band is the only one in which *mobile* transmission is permitted by the Federal Radio Commission without annoying restrictions.

Of the antenna systems illustrated here, the simplest 5-meter antenna system for a car or boat is a four foot vertical rod of stiff metal placed as far away from metal bodies as possible. In telescopic form, it can be easily dismantled as shown in the drawing.

The feeder system for any of these antennas consists of twisted lamp cord of the necessary length, connected to the antenna coupling coil in the transceiver.

Breastplate microphones such as are used in airplanes will be found practical for auto use, since they leave both hands free for driving. Modern transceivers will operate a magnetic speaker on good signals; this may be mounted under the dash.

Remember that every foot of height above ground increases the range of 5-meter transmission. In working from ground to plane, or where one station is located on a mountain, transmissions over distances of up to 100 miles are perfectly possible with a good transceiver.

# New Ideas for the RADIO EXPERIMENTER

**FIREPLACE SPEAKER** is ideal for homes having unused fireplaces. Get a sheet of ½" wallboard or plyboard that will just fit in fireplace opening. Cut a circular hole in center, to a size a bit smaller than rim of cone speaker. Mount speaker on panel, and run wires out through small hole in bottom corner of board. Glue grille cloth over opening in panel to improve appearance.

**SOLDERING IRON** will heat more rapidly in this stand, and may be used without troublesome cord. Fasten a plug to handle end of iron, and connect to heating element leads.

Connect remainder of cord to small outlet which can be fastened to workbench.

To use, plug in outlet cord, and insert iron.

The heat will rapidly rise to point of iron, where it is wanted, with iron in vertical position.

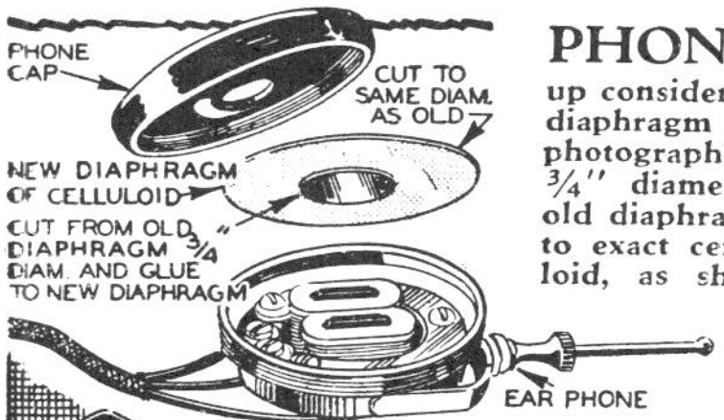


**PHONES** can be pepped up considerably by using diaphragm cut from old photographic film. Cut ¾" diameter disc from old diaphragm, and glue to exact center of celluloid, as shown at left.

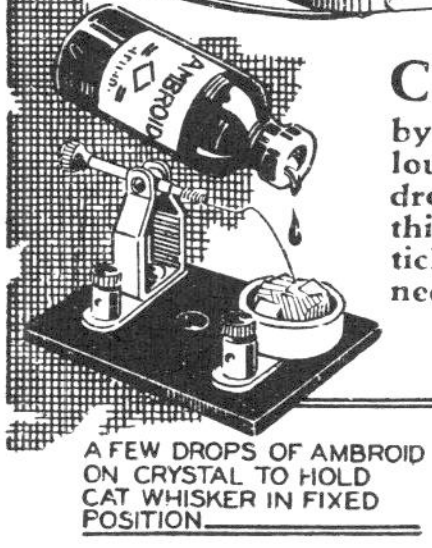


**CRYSTAL** detector can be "fixed for good" by adjusting cat whisker to the loudest point, and placing a few drops of Ambroid cement around this point. When dry no more tickling of the crystal will be necessary.

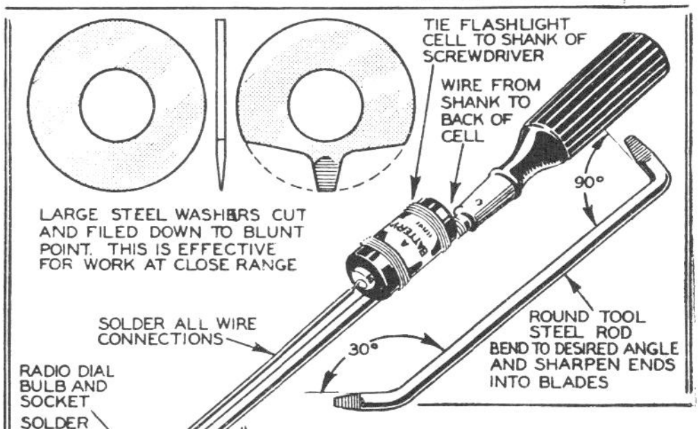


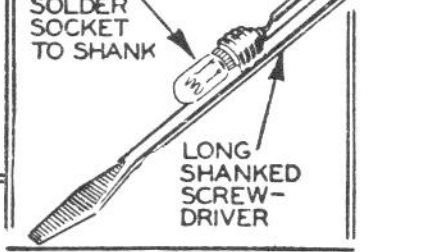


**SCREW-DRIVERS** for those odd corners are simple and easily made. Any steel washer filed down will do for work at close range, giving ample leverage.

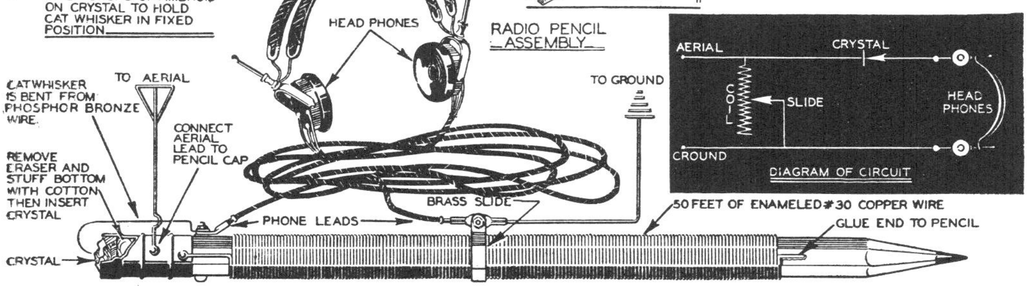


**CRYSTAL SET** designed by T. A. Blanchard is built on ⅜" diameter novelty pencil. Solder the top lead of the coil to the metal cap, and simply glue bottom end to the pencil. Remove eraser, push wad of cotton into bottom of cap, and insert a mounted crystal. Bend cat whisker from a length of phosphor bronze wire, and hold in place with bolt going through wood part of pencil. Scrape the enamel off a narrow strip of the coil, so that slide makes contact.

To use, connect aerial, ground, and headphones, move slider and adjust crystal until stations are heard.

# Build this "Double-M Special"

Fig. 1. Drawing suggests convenient layout for transmitter.

Fig. 2. All parts for the "Double-M Special" transmitter are mounted breadboard style as illustrated in this photo. Tuning is done with the variable condenser and feeder clip which attaches to copper tubing. Note position of R.F. choke.

By THORNTON HALLETT and R. I. CROWELL

**This little "Double-M Special" radio transmitter may look simple, but it's extremely efficient. Get on the air with it and you'll be amazed at the distance you can cover—and without the least trouble in getting it to "perk."**

EVER since the early days of radio transmission by amateurs, simplicity in transmitter construction has been the goal and now here is a low power transmitter to meet the demand.

Anyone with the slightest ability to construct radio apparatus can build and operate this transmitter because practically all of the "bugs" found in other outfits have been killed off.

### Advantages of the "Double-M Special"

Simplicity is the keynote. Single dial tuning, minimum number of parts, persistent oscillation at any point of tuning, high efficiency at low power and a stable frequency output to meet all the strict requirements now demanded by the Radio Commission—all these characteristics mark the "Double-M Special" transmitter.

For low-power operation this oscillator can be built entirely of receiver parts and so few of them that the cost, even at retail prices, will not exceed ten dollars. If parts are taken from discarded receivers, as were those in the one shown in the accompanying photographs, the cost will not exceed more than two dollars, exclusive of the tube.

This cost will be principally for the copper tubing used in winding the coil and for the coil plugs and fittings. The breadboard style shown is simple to put together and allows parts to be spread around and arranged as desired.

So, let's go, and build up the outfit and get it perking.

The baseboard size is unimportant but for the outfit shown it measures 10 inches long and 8 inches deep. Lay out the parts about as indicated in the diagram and base-

# 3-Band Code Transmitter

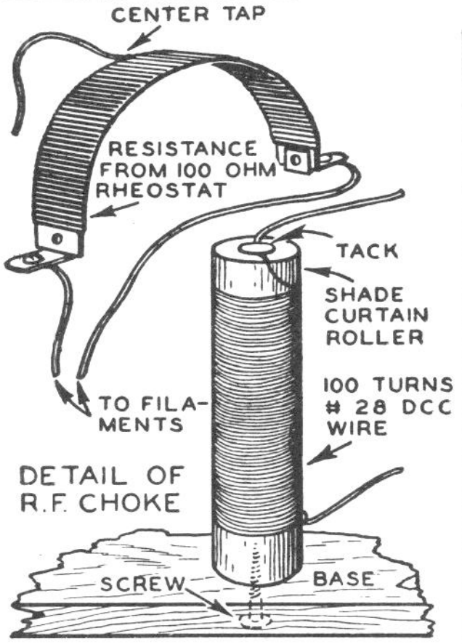


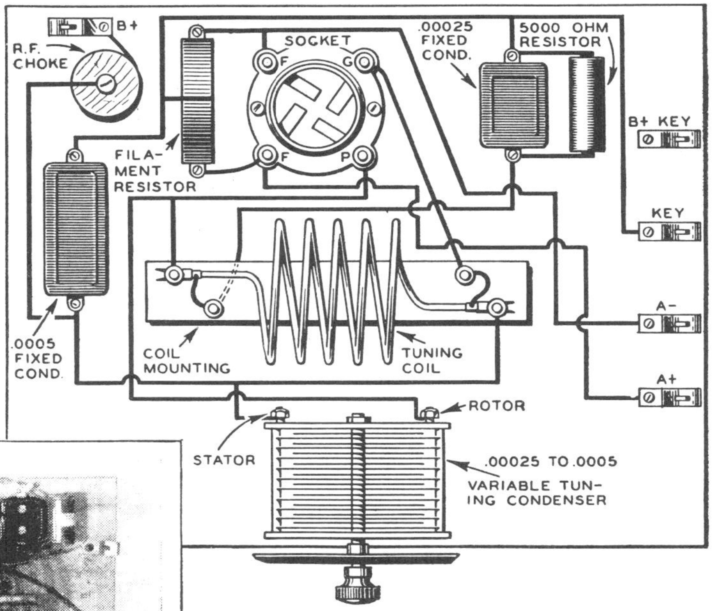


Fig. 4. Parts mounted on baseboard are wired up as illustrated here. Note particularly arrangement of leads from the grid coil. Upper left—Construction of R. F. choke and resistor, made from old filament rheostat.

Fig. 3. The top view of the transmitter illustrates clearly the arrangement of all parts on the baseboard. An extra Fahnestock clip is placed on side for the sending key.

board sketch. Place the socket in the rear center, the radio frequency choke to its left (vertically) as well as the plate blocking condenser. See Figs. 2 and 4.

At the right of the socket place the grid leak and condenser. In the center front mount the variable tuning condenser and between it and the socket place the plug-in base for the coil. Four connecting clips are fastened near the right hand edge for power connections as well as one clip at the rear of the choke.

### Here's All Apparatus You Need

The apparatus required for 20, 40 and 80 meter operation is as follows.

- Plate coil of 1/8 inch soft copper tubing.
- Grid coil of No. 18 rubber insulated wire.
- Variable tuning condenser. Can be from .0005 to .00025 capacity.
- .0005 fixed condenser. Minimum breakdown rating not less than 500 volts.
- .00025 grid condensers.
- Two .002 fixed condensers.
- Filament center tap resistors. (These can be 100 to 200 ohms each side of center tap, or Xmas tree lamps)
- 5000 ohms grid leak (5 watt).
- Radio Frequency choke. About 100 turns No. 28 to No. 30 DCC wire on 1 inch form.
- One UX or UV socket.
- Four Fahnestock clips.
- Four General Radio plug mountings and sockets.
- Hook-up wire.
- Thin wood 1 inch wide.
- Screws, bolts etc. for assembling.

### Secret of Efficiency in Coils

Now the whole secret of this oscillator lies in the construction of the coil alone. Look at the photos and sketches and you will see but one tuning coil, yet there are two. That is the whole idea of the "Double-M Special" in a nutshell.

The diagram (Fig. 6) shows two coils, the plate and grid coils, but in Fig. 4 you will notice only one. The plate coil is apparently tuned, whereas in the usual Tuned-

**Fig. 5. Rear view of transmitter above shows clearly position of R. F. choke, tube, resistors, and plug-in coil base.**

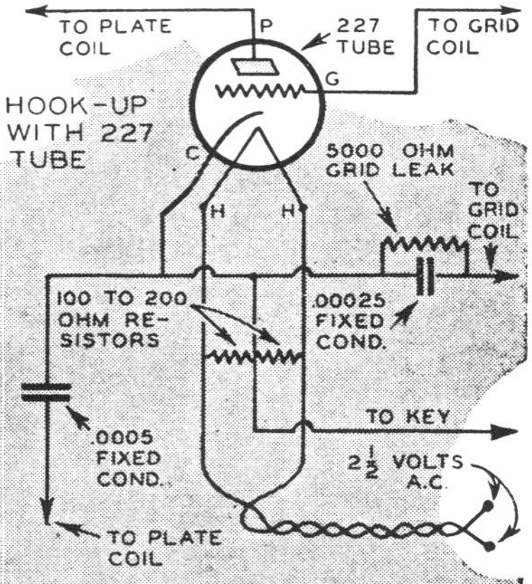


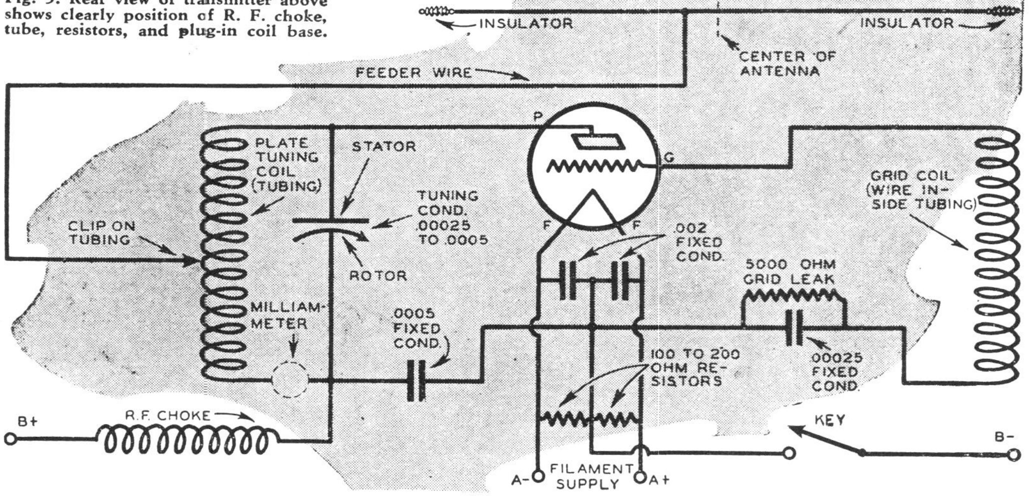


**Fig. 6. Here is the complete wiring diagram of the "Double-M Special." Feeder wire from antenna attaches with clip to plate coil (copper tubing) inside which is wound the grid coil (wire). Milliammeter is optional. Top right—Hook-up for 227 tube. In tuning, a Christmas tree lamp can be inserted in feeder lead to indicate resonance.**

Plate Tuned-Grid circuit both plate and grid coils have separate tuning condensers to keep them in resonance.

Now in this circuit these coils are kept in constant resonance at any frequency by being so tightly coupled that they can't help being resonant at all times, and the tuning of one automatically tunes the other. There is no chance for one circuit to outbalance the other and "slop over", thus overheating the tube or doing other tricky stunts to throw the whole transmitter out of whack.

Feedback is always present and varies not a whit with the tuning. This close coupling is accomplished by winding one coil inside the other!

But, remember that the coupling *must be tight*. That is, the size of one coil must be almost an exact duplicate of the other, as near as mechanically possible.

Therefore use wire for the grid coil that will just squeeze through the tubing composing the plate coil. Using small wire that fits loosely in the tubing will not do. In the coil shown in Fig. 7 wire, insulated with rubber for outside use, was pushed through the straight tubing only by squirting a small quantity of oil in the tubing and then forcing the wire through inch by inch.

But it made a perfect job. Fig. 7 shows how the coil looked after it was wound, with plenty of the wire projecting from each end to attach later to the fittings.

After the wire is through and the coil wound approximately 3 inches in diameter

# *Interchangeable Coils Permit Quick Shift 20, 40 or 80 Meter Wave Bands*

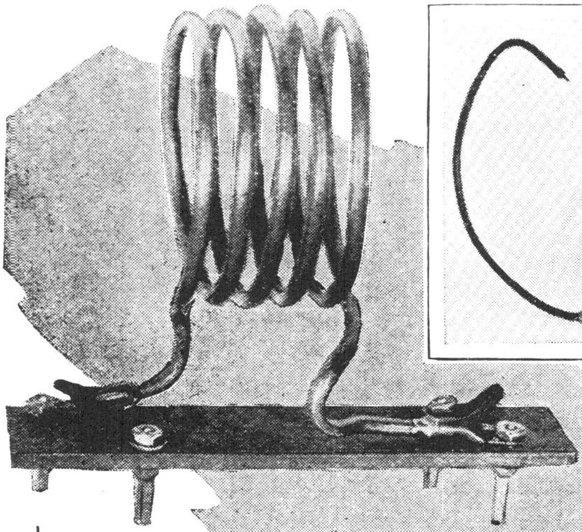

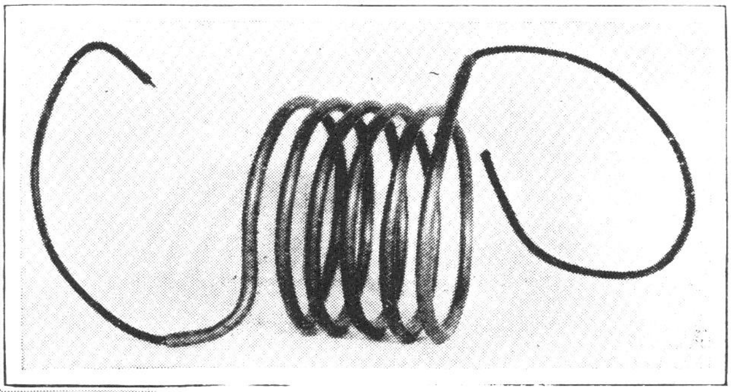

Fig. 7. Wire of grid coil is wound in copper tubing.

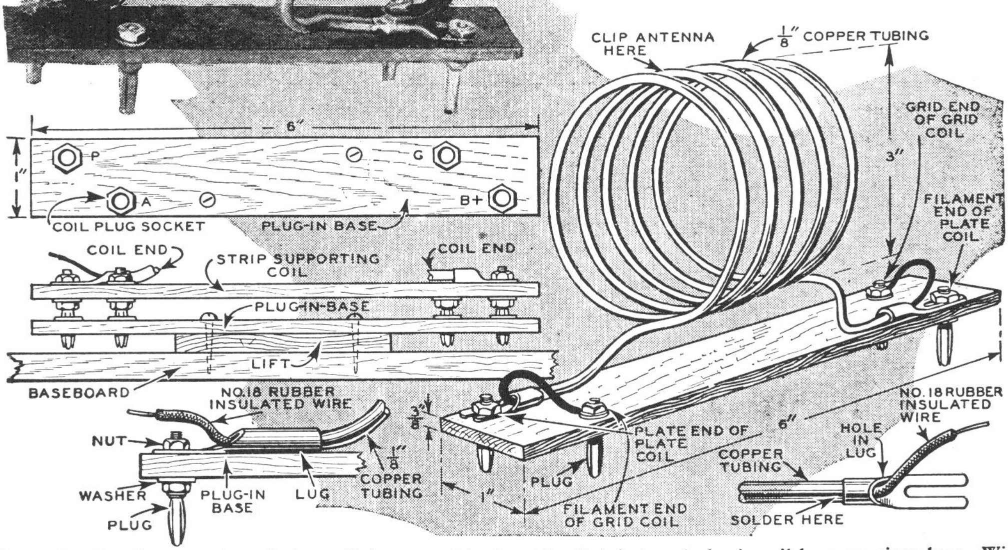


Fig. 8. Details of construction of plate coil (copper tubing), grid coil (wire) and plug-in coil base are given here. Wire is led into copper tubing through hole bored in soldering lug. In winding, take care not to damage insulation of wire.

around an old dry cell, the ends of each coil must then be fitted to the mount.

Study the detail diagrams of the coils in Fig. 8 closely. You will note a lug soldered to each end of the coil in a horizontal position. These can be made from short pieces of ¼ inch tubing pounded flat half their length and sawed up in a slot.

Drill a hole through the round part of the lugs, push the wire ends through the holes and then force the lugs over the ends of the tubing and solder in position.

Now cut off one of the thin wood strips about 6 inches long and fit it with the four radio plugs as shown in Fig. 8. Secure the left end of copper tubing to the end plug. This is the plate end. Then bolt the right end lug to that end plug which is the B plus plug.

The wire coil is then bolted to the two remaining plugs but the markings must be exactly as shown. That is the left one is marked "A minus" and the right one is marked "G" (grid).

It will be noted that this reverses the action of the two coils from the way they are shown in the diagram and is necessary to the proper working of the circuit.

**Constructing the Plug-in Base**

Between the tuning condenser and socket mount a similar strip of wood for the plug-in base which is raised from the baseboard by a short piece or lift. This is also shown in detail in Fig. 8.

The plug sockets must be fitted to this strip so the plugs will fit in them nicely when pushed down. Remove the coil and you are ready to wire the outfit up.

First mark the four clips at the right of base as shown—A plus, A minus, Key—B minus and Key. The one at the R. F. choke is B plus. From A plus wire to one filament

*(Continued on page 20)*

# Tighten Coupling When Short Wave Coils Won't Oscillate

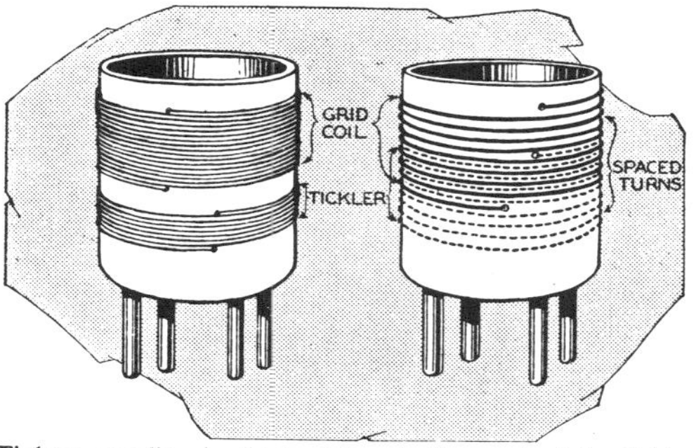


Tighter coupling is obtained by winding part of the tickler coil between the turns of the grid coils as shown above.

MANY, many times builders of short wave receivers wind their own coils and are then unable to make the set oscillate. It is the coils wound for the higher frequencies that furnish the most trouble.

Now the trouble is often a matter of too loose coupling or too few tickler turns. One excellent way to remedy this is to wind a portion of the tickler turns *between* the grid turns. For very high frequencies, say up around 14,000 to 28,000, the entire tickler can be wound inside the grid windings with excellent results whereas with the two windings separate the set would not oscillate at all.

# "Double M Special" One-Tube Three Band Code Transmitter

*(Continued from page 19)*

of socket and to one end of center tap resistor.

Ordinarily, you will need separate coils for each amateur band, but the number of turns will be found to be less than is usually the case with a Hartley or other common circuit. Thus in the transmitter shown a 10 turn coil sufficed nicely for 80 meters, 5 or 6 turns for 40 meters and 3 turns for 20 meters using a .0005 tuning condenser.

### How to Compute Tubing Length

To find the length of tubing needed multiply the desired diameter by 3.1416 and then by the number of turns and allow about 6 inches more for ends. Thus for an 80 meter coil you will need about 100 inches or 8 ft. 4 inches of tubing.

The size tubing and wire specified here is suitable for tubes up to the 210 size and not over 300 volts of plate power. For 210 and larger use ¼-inch tubing and No. 14 wire.

In any case, when building the coils make very sure the insulation on the grid coil is heavy, not broken or punctured at any point, and that where it emerges from the copper tubing (plate coil) that it does not become cut or cracked.

Just as sure as this happens the R. F. currents in the grid coil will short to the plate coil and oscillation will cease. But careful construction of the coils with these hazards in mind will assure you a good safe job that will surely work.

This outfit as shown is intended to be used with a voltage fed Hertz antenna and the feeder clips directly to the plate end at about the second turn from end P on the 80 and 40 meter coils and on the first turn for 20 meters.

Other types of antenna can be used if desired but call for a second antenna coil mounted an inch or two away from the plate end of the oscillating coil. For operation on 80 meter the writer uses an antenna 133 feet long, with the feeder wire attached 18 ft. 4 in. from the center, as shown in Fig. 6.

For the most efficient operation on low power, a 180 volt B eliminator can be used, with an A battery to light the tube. Simply connect the A power to A plus and A minus clips, B plus of the eliminator to B plus clip and the B minus wire to the B minus and Key clip.

Then connect the key to the two clips marked Key. With the tube lighted and the eliminator plugged in, press the key and the tube will oscillate.

### Test for Oscillation With Flashlight Bulb

To test for oscillation, make a single loop of wire connected to a flashlight bulb and clip it a short distance away from the plate end of the coil. When the key is pressed the bulb will glow.

Then the frequency can be set within the band by the use of a wave meter as described in previous articles. When you have it perking simply clip on the antenna feeder, retune slightly and you are ready to send.

The little detail sketch (Fig. 4) also shows how a filament center tap resistance can be easily made from an old 100 ohm rheostat winding and solves that problem in a simple manner. Also, if an AC tube (227) is to be used for an oscillator, a diagram (upper drawing, Fig. 6) is attached showing how this should be wired in.

The heater terminals are wired direct to a 2.5 volt source and center tapped as before but the 2.0mfd. fixed condensers are not so necessary. In fact they can be dispensed with anyway if the other type of tubes are lighted with DC current.

# An Easily Made "BUG" for HIGH SPEED Code Work

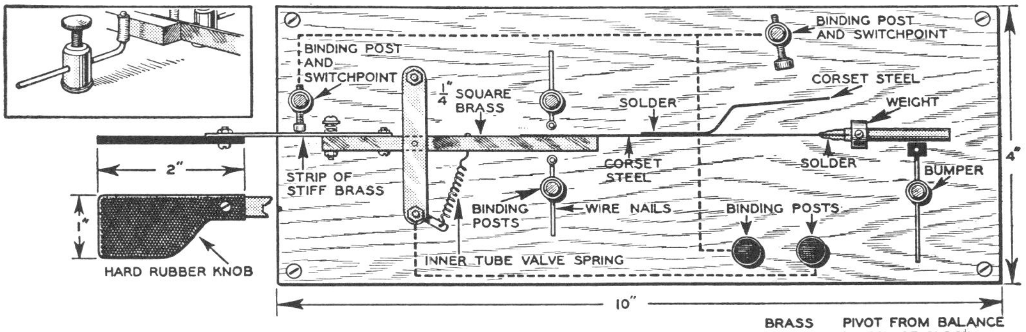


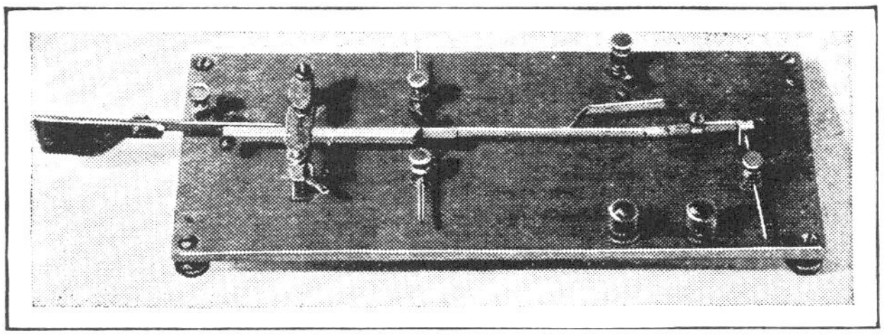

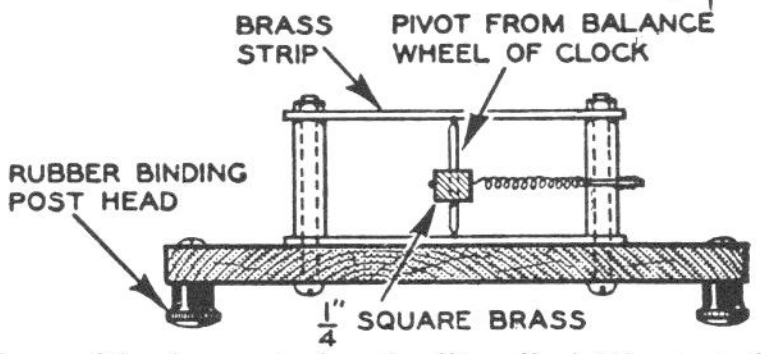


**Assemble the parts for the "bug" as illustrated in the drawings here. Throwing the knob to the left makes a dash, while throwing to right makes a series of dots, speed depending on springiness of steel and weights on end.**

WHEN a man becomes really enthusiastic about the amateur radio transmitting game he yearns for a high speed key for rapid fire operating.

The bug shown in the photograph was built entirely of spare junk around the shop and works just about as nice as could be desired.

The base consists of a piece of thin wood 10 in. long by 4 in. wide mounted on four "feet" consisting of a hard rubber binding post head under each corner. The pivoted arm consists of four parts. The main portion is a piece of ¼ in. square brass about 4 in. long. To one end is soldered a piece of narrow corset steel 4 in. long and to the other end is bolted a strip of stiff brass to project beyond the end about 1½ inch.

**Steel Spring Gives Vibrating Action**

A second piece of corset steel should be bent in the form indicated and soldered to the first in about the location shown. Now to the end of the brass strip bolt a knob consisting of a piece of bakelite or hard rubber cut in the shape shown in detail. At the rear end solder some sort of weight. In the bug shown this weight consisted of a short piece of variable condenser shaft fastened on with a collar and set screw.

The pivot frame was taken from an old alarm clock but any similar affair can be used. The pivot was that supporting the balance wheel. A small hole was drilled through the brass piece about a third way down its length from the rear end and the pivot forced through and soldered. The pivot was then mounted in its frame and the latter fastened to the base in the position shown. Tension of the pivot bearings was then adjusted so the arm swung freely but without lost motion.

**Making the Bumpers**

Now near knob and contacting the brass strip is placed a binding post into which is adjusted a switchpoint as shown. This contact occurs when the brass bar is centered between the bumpers. A second binding post and contact is placed opposite the projecting corset spring so it normally rests about ½ inch back. Then a rear bumper is placed near the end of weight in the position shown.

This bumper can be made as the two bumpers ahead are made, or instead of spaghetti a small piece of eraser can be speared onto the upright part of the nail. Lastly connect an inner tube valve spring in the position indicated. This spring tension must be regulated to obtain the proper working tension.

When completed the bug can be hooked into the key circuit or shunted across it as an auxiliary key. Its action is as follows: When the knob is pressed to the left so the brass strip contacts its binding post the circuit is completed and a dash is made as long as the contact is maintained. Releasing the knob the spring carries the pivoted arm over and the piece of corset steel vibrates against contact, the rear contact making dots at each contact. This rate of vibration is greatly determined by the size of the weight and the flexibility of the spring, which you can adjust to suit.

# Build Short Wave Adapter

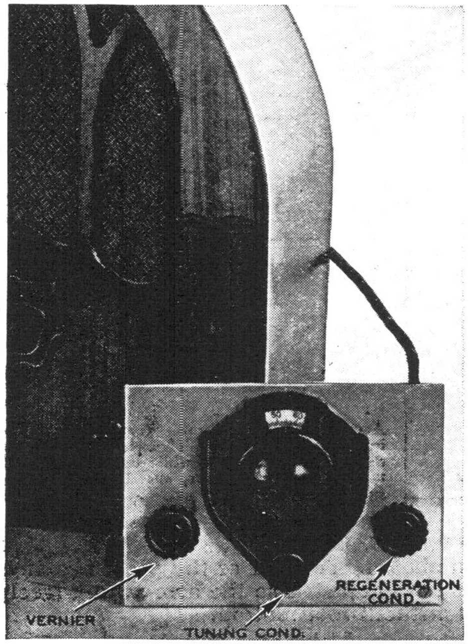


Short wave broadcasts, police stations, and amateurs are tuned in with the tuning condenser (center) and regeneration condenser (right). Make fine adjustments with vernier.

by B. A. THOMAS

DAILY the number of short wave listeners increases. It will be but a short time until the number of listeners on the lower wavelengths will be as great as the number of listeners on the broadcast band when the DX craze was at its peak several years ago.

Although this information is not at all new to many short wave fans, they may nevertheless profit by it, and get a kick out of it by building this little adapter.

If your receiver has a hefty push-pull output you can expect to have a short wave receiver with plenty of pep. This little device, when used in conjunction with a broadcast receiver having but two tubes in the A. F., brought in all of the domestic shortwave transmissions, as well as five South and Central American stations and a host of European stations.

Police, ship to shore, airplane and ham transmissions are so numerous that they are passed by, but for the newcomer on shortwaves they offer plenty of excitement. If you have never had the privilege of listening to broadcasts which are really interesting and in most cases free from advertising, this adapter is what you want.

Arrange parts according to layout shown here, making all connections with bus bar wiring and keeping the leads short.

# for Your Broadcast RADIO

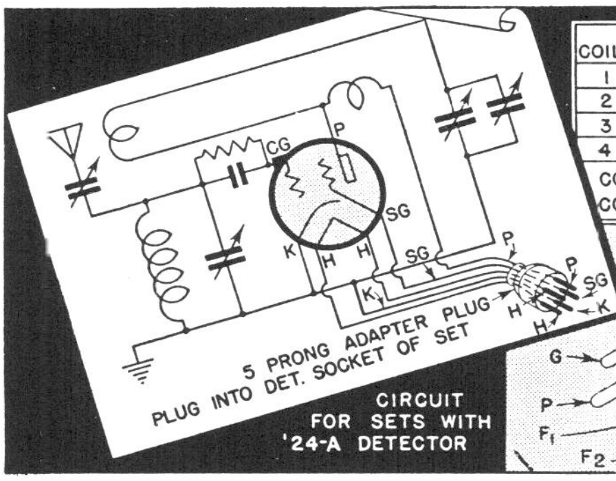


| COIL | METERS | $L_1$ TURNS | $L_2$ TURNS |
|---|---|---|---|
| 1 | 15 - 25 | 5 T. | 7 T. |
| 2 | 25 - 49 | 12 T. | 9 T. |
| 3 | 49 - 105 | 24 T. | 12 T. |
| 4 | 105 - 200 | 50 T. | 15 T. |

COILS 1,2,3 — 24 D.C.C.
COIL 4 — 28 ENAM.

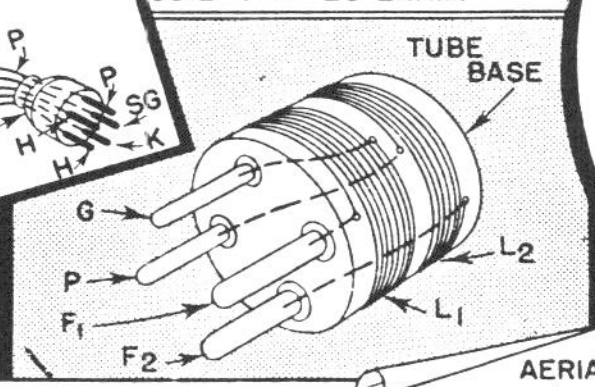


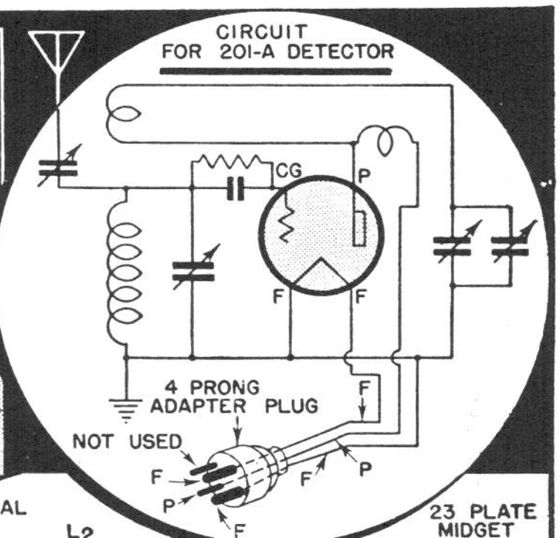


Plug and tube connections for sets having '01A and '24A detector tubes are shown above, together with coil winding chart and coil form connections. Shellac windings to prevent slipping of turns. Circuit constants are in diagram below.

After drilling the aluminum panel you can mount the three variable midget condensers in their respective places, as illustrated in an accompanying photo. On the right is mounted the 23 plate regeneration or throttle condenser. On the left a 3 plate vernier regeneration control (optional) and in the center another 23 plate type for tuning control.

You will notice, after examination of the circuit, that the condensers are not insulated from the panel. Since the panel is in the cathode-gnd. circuit and the rotors of all three control condensers are also in this circuit, the panel acts as the connection to the three.

The leads emanating from the adapter go to an adapter plug which in turn is plugged into the detector socket of the receiver. When the adapter is ready for use plug in and connect the receiver antenna to the antenna post on the adapter.

Signals are tuned in on the adapter controls and the receiver controls are "dead" with the exception sometimes of the volume control.

Turn the knob on the 23 plate throttle condenser until a "plop" is heard. Stop turning and retard the control until the receiver is on the verge of "plopping" again. At this point the tuning control is turned until a heterodyne whistle is picked up or the actual music or voice.

After some practice with the adapter you will be able to clear up the whistle and get the voice or music just as clear as on your regular receiver.

The R.F.C. (radio freq. choke) shown is a home-made affair and consists of 350 turns of any fine wire (insulated, of course,) wound on a short wooden dowel about ⅜" in diameter. The grid leak and condenser

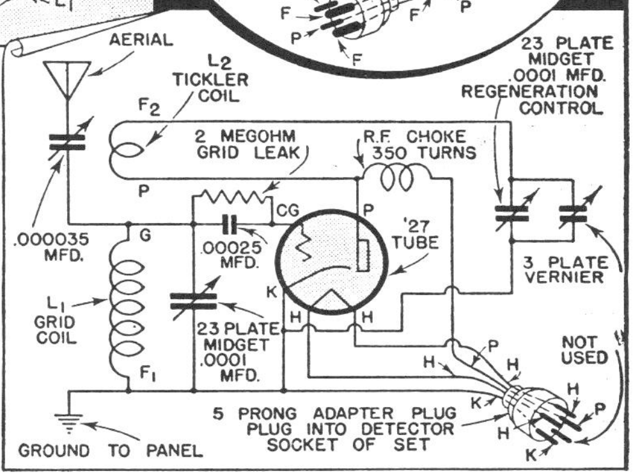


Circuit diagram of adapter for set having '27 detector tube. Remove detector tube from set, insert in adapter socket, and place plug in vacant socket in receiver. Antenna condenser is trimmer or bookplate type, adjusted with fiber screwdriver.

are mounted between the two sockets and the leads are made as short as possible.

The best detector tube to use is an Arcturus type 127. The newer types of other brand tubes will do, but none quite as good as this. Sets using a –37 or –56 detector tube will also work with the original adapter circuit. For other type detector tubes the circuit must be changed slightly, as shown in the sketches for battery and screen grid sets.

Best results will be obtained on the earlier model sets. The adapter will work on any tuned radio frequency set having a good audio frequency circuit; on other radios use the converter.

If the set does not regenerate the trouble is either due to the coils being wound in opposition directions, or because the connections to coil No. 2, the plate coil, were reversed.

The 23 plate midget tuning condenser, with a maximum capacity of .0001 mfd., is used for both adapter and converter circuits. A 13 plate midget will not be satisfactory unless more turns are used on the coils.

# Changing the S. W. Adapter to a Converter for Any Set

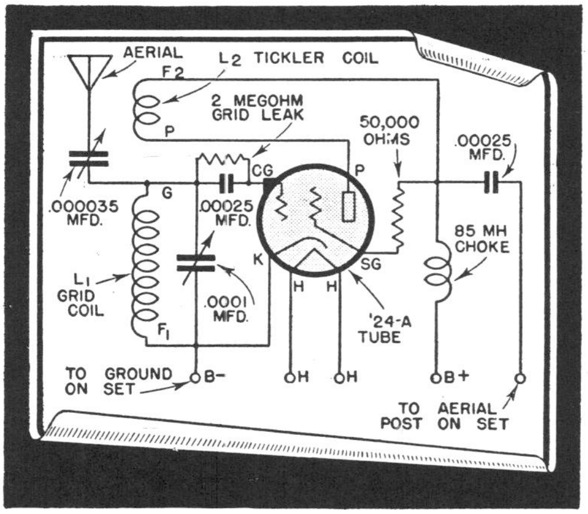


**Adapter parts are rewired as converter in above diagram.**

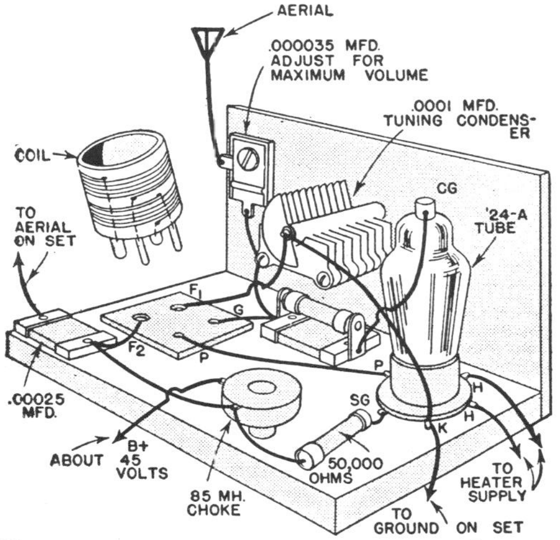


**Pictorial diagram of converter gives one arrangement of parts which has worked out very well in tests. Adjust antenna trimmer condenser for maximum volume when changing coils, or set at point of fair reception with all coils.**

BY CHANGING a few connections in the short wave adapter circuit shown on page 23 an efficient converter which will work on any set, battery or electric, superheterodyne or tuned radio frequency, is obtained. This will, if properly assembled, be far superior to the adapter, bringing in police, amateur and even foreign broadcasts through the loudspeaker of your own radio set.

With this converter all of the tubes in your receiver are functioning, whereas in an adapter you are using only the audio stages for short wave reception.

**No Regenerative Squeals Here**

The converter tunes with but one control, eliminating the bothersome regenerative squeals of the adapter. A separate type '24A tube is used in the converter.

In rewiring the adapter the five prong socket is used as before. Now, however, the G prong on the socket is for the screen grid connection of the '24A tube. Connections for the control grid are made to the cap at the top of the tube.

The filament of the tube requires 2.5 volts at 1.75 amperes, either alternating or direct current. A 2 ohm resistor placed in series with a 6-volt storage battery will give you the correct voltage, or a separate filament transformer having a 2.5 volt secondary and 110 volt primary may be used. If there are any tubes in the receiver using 2.5 volts, the filament supply for the converter may be obtained from one of them.

**Apply About 45 Volts to Plate**

A plate voltage of somewhere around 45 volts will be best for the converter. This can be obtained from a 45 volt B battery, or taken from the receiver itself. On some sets the correct voltage can be obtained from the SG prong of the first r.f. tube. It is best to measure this voltage with a direct current meter first, to be sure you are getting the correct voltage from the set.

The B– connection is obtained by running a wire from the ground terminal on converter to ground terminal on receiver, and grounding this wire at either point.

If finer tuning is desired, a three plate vernier condenser may be connected in parallel with the large tuning condenser.

The coils used here are exactly the same as those used in the original adapter. If you are in doubt as to whether you have wound the coils correctly, try reversing the grid coil leads.

In any converter circuit the performance of the receiver is just as important as the design of the converter. If your set will not pick up distant stations with plenty of volume on the broadcast band, you cannot expect foreign reception in short wave bands.

## Handy Lamp Socket Resistor Bank

A HEAVY duty variable resistance will often save many fuses when testing electric motors. An arrangement of ordinary lamp sockets and knife switches will provide a convenient variable load resistance. Mount three or more lamp sockets and knife switches on a baseboard, and connect so either series or parallel combinations can be obtained.

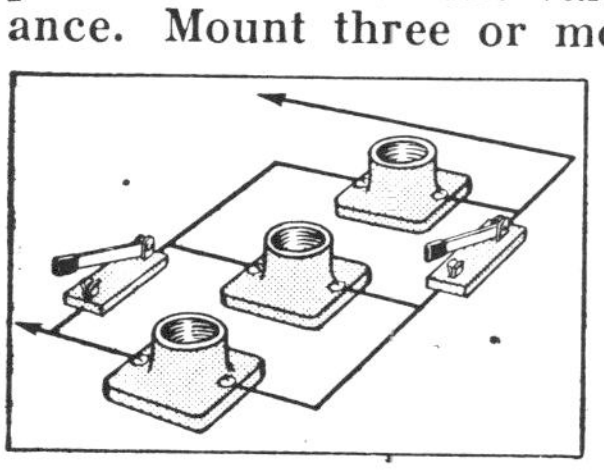

Lamp bulbs of different sizes may be used in the sockets. For heavier loads use resistance heater coils.

# Making a Doublet Antenna for Receiving, Transmitting

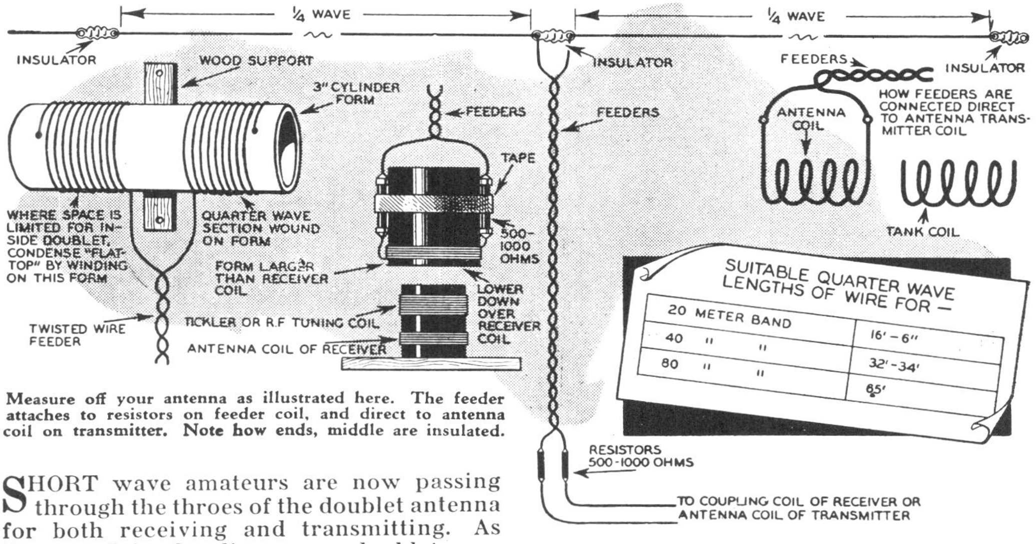


| | |
|---|---|
| 20 METER BAND | 16'-6" |
| 40 " " | 32'-34' |
| 80 " " | 65' |

**Measure off your antenna as illustrated here. The feeder attaches to resistors on feeder coil, and direct to antenna coil on transmitter. Note how ends, middle are insulated.**

SHORT wave amateurs are now passing through the throes of the doublet antenna for both receiving and transmitting. As illustrated in the diagram, a doublet consists of a "half wave" flat top cut in the exact center and fed by two feeders. The proportions of the flat top are always the same but the feeder lengths can vary by odd quarter wavelengths.

For example, we will take the doublet for operating in the center of the 20 meter amateur band. Using any good antenna wire, cut two pieces 16 ft. 6 in. long. These are quarter wave long for that frequency of 14200 kilocycles. Connect them together by an insulator and, of course, put an insulator at each end. This constitutes the flat top.

Then, to suit the distance from the antenna to the set, make the feeders of twisted lamp cord either one, three, five or more quarter wavelengths long. In this case it will mean either 16 ft. 6 in., 49 ft. 6 in., etc., long.

Connect the two wires at one end of the cord to the inside ends of the flat top as shown, soldering the joints, and connect the other two ends to the set as shown in detail. Lengths for other waves are shown in the table.

Coupling to a receiver can be best done as shown in the sketch. This consists of a coupling coil of about half a dozen turns of No. 20 to No. 24 DCC wire wound on a form that will just slip down over the antenna tuning coil of the receiver so this coil lies from an eighth to a quarter of an inch away from the antenna coil itself.

Strap two 500 or 1000 ohm resistors outside the form and connect the leads of the coupling coil to one end of the resistors and the feeder ends to the other ends of the resistors. That is all; the receiver tuning is done as usual.

To couple to a transmitter it is only necessary to attach the feeder ends to the ends of the antenna coupling coil. Some amateurs make a practice of tapping each feeder one or two turns in from the end of the coil, claiming it clears up the note. As the feeders are a quarter wave long no feeder tuning is needed as with a Zepplin type antenna.

Now for the listener with small quarters. If you haven't room to spread out a doublet you can make it very compact by winding the flat top portion on a 3-inch form as shown in the drawing. Just mount the cylinder on a suitable stick. Then wind each portion of the flat top (in the same direction) on the form about two inches apart and connect on the feeders. This works very nicely where room is at a premium and is intended for indoor use.

The doublet appears to be a very efficient antenna system and its simplicity makes it worth while building. Its chief claim is the low noise level to signal ratio.

## Insulate Wire in Awkward Places

WHEN you encounter the task of insulating a piece of bus bar at an inconvenient place in your radio set, first slit the piece of spaghetti tubing, then slip it over the wire, and shellac as shown in the accompanying drawing.

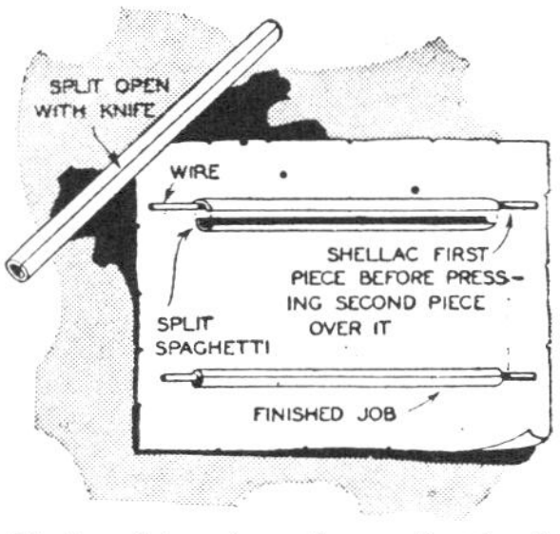


**To insulate wire, slit spaghetti tubing, slip over wire, and shellac. This forms an insulated conductor well down in receiver's innards.**

# Ordinary Voltmeter Measures Filter Condenser Capacities

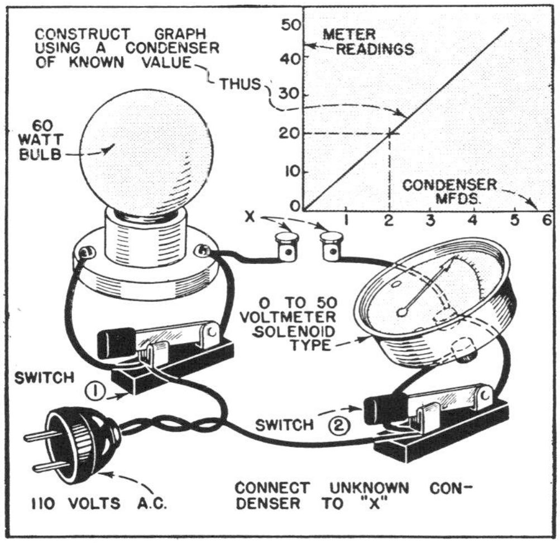


Mount meter, bulb, and switches on test panel. Light bulb gives indication of condition of condenser; meter gives rating. Use 0-50 volt, solenoid type B-battery testing meter.

OPERATING directly from 110 volts, a.c., this simple circuit will test and measure quickly the capacity of radio filter condensers.

To calibrate the tester, connect a condenser of known capacity to the X posts, close switch 1, open switch 2, and observe the meter reading. Plot this reading on a graph, then draw a straight line through the point and the zero of the graph.

In testing unknown condensers, close switch 2 and open 1. If the bulb glows brightly, the condenser is shorted. If it remains dark, glows faintly, or flashes for an instant, open switch 2, close 1, and note the meter reading. Pick off the corresponding condenser rating from the graph.

## Gadgets Make Wire-handling Easy

WHERE a multitude of bends are to be made in stiff wire, a bending jig will prove a time-saver. By hinging two hardwood blocks together, cutting a notch corresponding to the desired shape of bend in one, and a projection of the same shape on the other, an inexpensive but handy jig can be made. It will be found useful especially for bending bus-bar wire for radio sets.

Heavy wire is pulled taut readily if gripped with a holder made from an old auto brake rod. Run the wire through the loop, then wrap it two or three times around the shank.

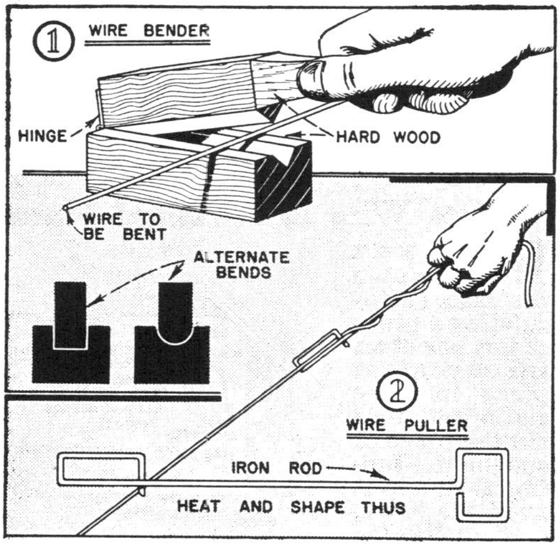


Fig. 1 shows wire bender in use. Design is readily changed if other shape bends are wanted. Wire puller in Fig. 2 is forged from old auto brake rod; heat before bending.

## Control Voltage Without Rheostat

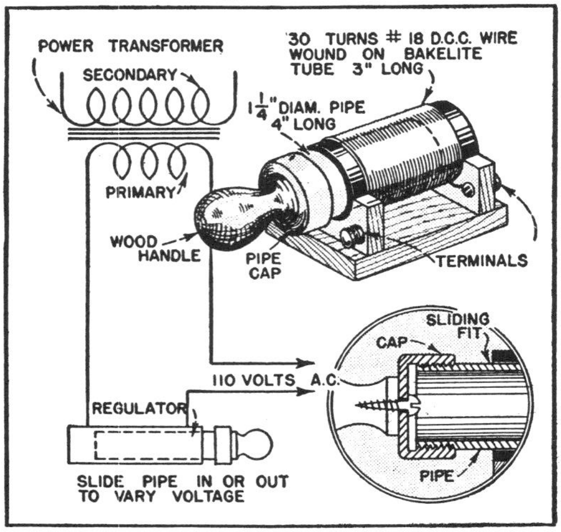


This voltage regulator for filament transformers is mainly choke coil connected in series with transformer primary, with adjustable iron core to vary effective a.c. inductance.

FILAMENT transformer voltages are lowered easily without a rheostat, by inserting this adjustable choke in series with the transformer primary. By changing the position of the plunger with relation to the coil, varying voltages can be obtained easily and quickly.

Wind 30 turns of No. 18 double cotton covered wire close together upon a length of 1½ inch diameter bakelite or cardboard tubing. Fasten the ends of the coil to binding posts mounted on a wood base made for the coil.

The plunger is a four inch length of iron pipe just large enough to slide inside the coil form. Force a wood handle into one end of the pipe, or bolt a handle to a pipe cap fitted over the end. The iron core changes the effective inductance of coil in proportion to distance it extends into coil.

# ENJOY RADIO MUSIC ON YOUR BICYCLE TOUR

Listen to radio music, amateur broadcasts, and keep in touch with world as you tour the country on a bicycle. Any small battery set using new 2-volt tubes can be used.

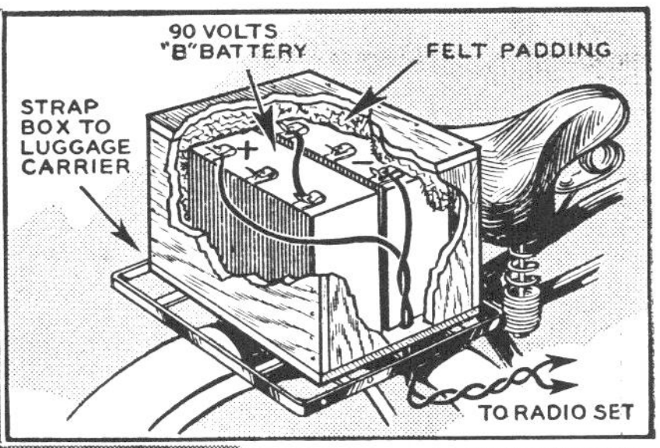


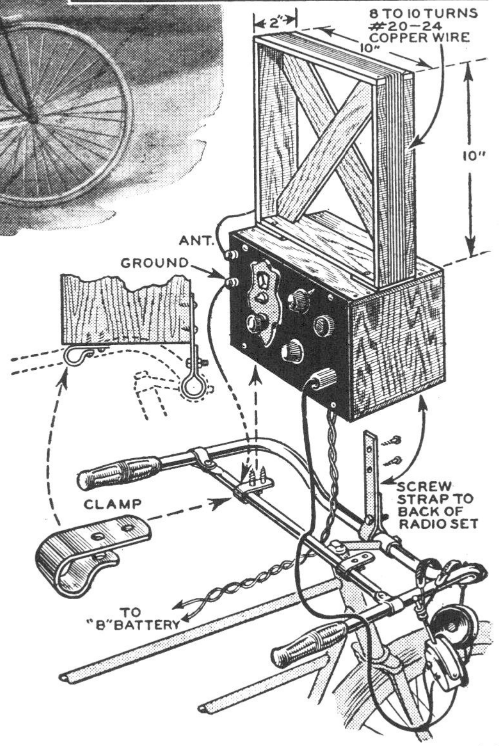


Frame of bicycle is used for radio ground. Loop aerial is mounted on top of radio cabinet, batteries on carrier.

CAR radios are now quite common, but few indeed are the radio-equipped bicycles. It really is a simple job to rig up the old wheel for radio, though you will have to be satisfied with headphone reception.

Either the Hitch Hiker's 2-tube Portable, on page 8 of this book, the 3-tube Tool Box Portable on page 104, or any other small battery radio receiver will fill the bill. The set should be small enough to fit between the handlebars, and preferably should give both long and short wave reception. The only change necessary in the sets will be to install cushioned tube sockets to take up road shocks. A square of sponge rubber slipped under the rigid sockets can be used if it is impractical to replace the sockets.

Make clamps of fairly stiff metal, to fasten the radio set to the handlebars. On bicycles without a crossbar the front clamps can be attached to the handlebars.

A loop antenna wound on a 10″ square frame of thin boards is used. Brace the inside of the square with cross members, and boil the entire framework in paraffin for half an hour. Wind 8 to 10 turns of No. 20 to No. 24 insulated wire on the loop, spacing out the turns a little. Paint the winding with melted paraffin, and bolt the antenna to the top of the radio set. One end of the winding is connected to the radio set aerial, the other left free.

Connect the radio ground to the metal part of the bicycle, such as a clamp.

Filament dry cells can either be mounted inside the radio set, or placed in the tool compartment of the bicycle.

The B batteries may be placed in a waterproofed wood box and strapped onto the carrier in back. Use rubber covered wire wound around the bicycle frame to make connections to the set. Stuff the box with packing to keep the batteries from moving around. Use a good pair of headphones. The volume will not be quite as great as when a long high aerial is used, but local programs will be more than loud enough.

All the more enjoyment will be obtained from a set powerful enough to operate a loudspeaker; here there will be no cord to get in the way as with the headphones shown above.

# How To Read RADIO CIRCUIT SYMBOLS

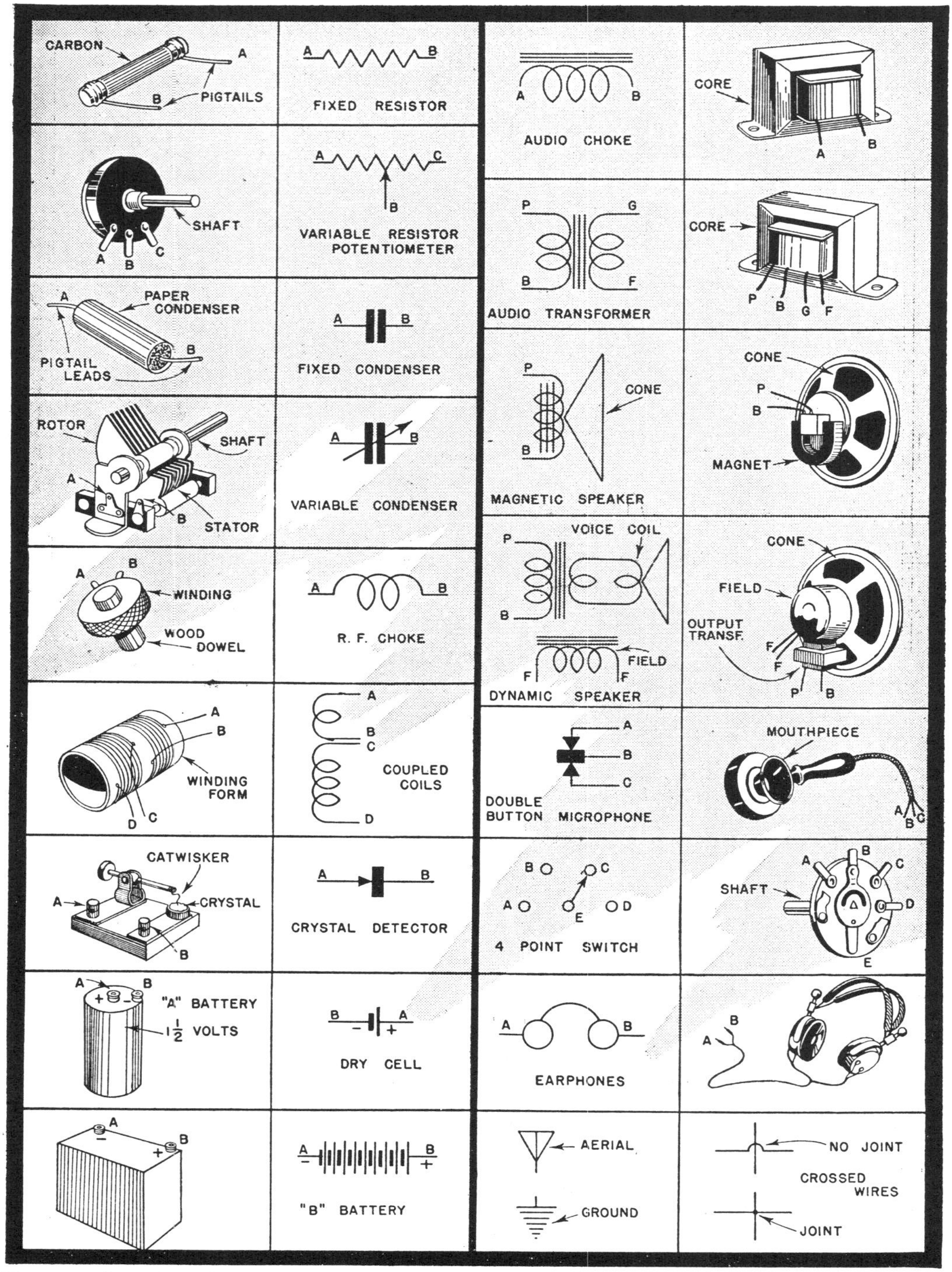


Once you understand meanings of schematic symbols pictured in two vertical center columns above, radio circuit diagrams will no longer be a mystery. Pictorial sketches are shown at right or left of each symbol, with letters to show proper connections. Remember that short bar on battery symbol is always negative. Use terminals A and B for variable resistor; all three for potentiometer. B on potentiometer is always movable contact. Parallel lines always indicate iron core.

# Making TUBE SOCKET CONNECTIONS

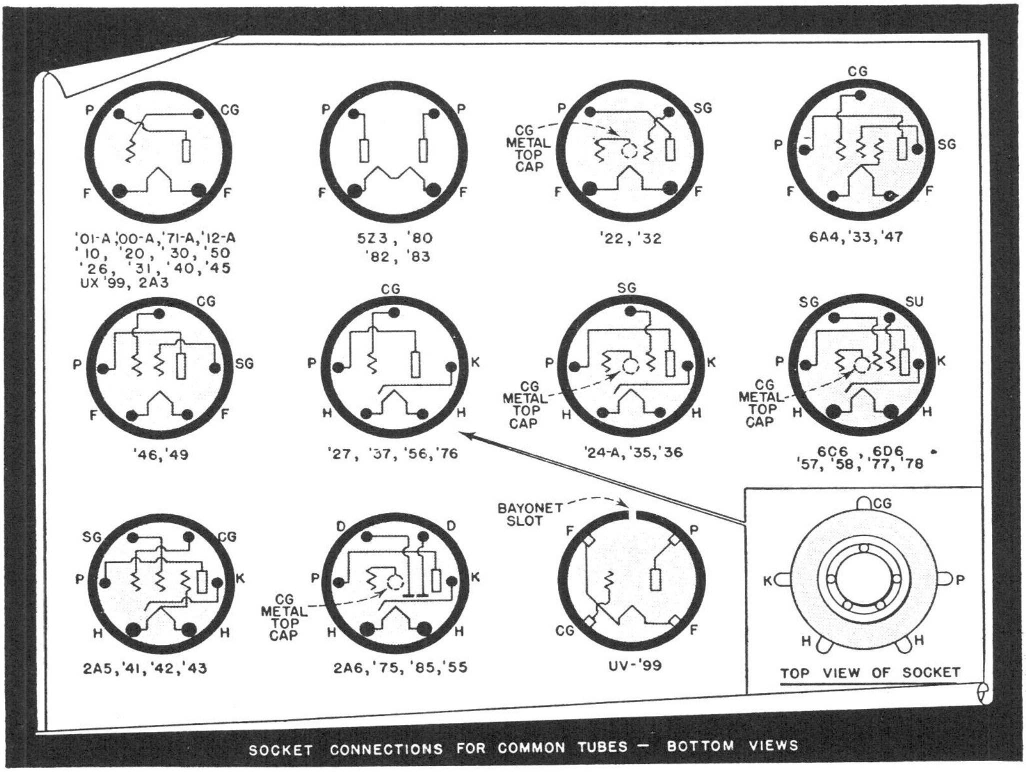


**Above chart shows connections for common types of vacuum tubes, as viewed from UNDER SIDE of socket, or as looking up at tube from prong end. When looking at top of socket, as in lower right sketch, connections are reversed.**

UNLESS a radio tube manual of some kind is available, it is rather difficult for the radio builder to locate proper connections for the different vacuum tubes specified in schematic circuits.

The above diagrams show connections for the more common types of vacuum tubes. Remember that you will have to reverse these diagrams, as if looking at them from the back of the page, when making connections to the socket from above.

Abbreviations used in the above sketch and in schematic radio circuit diagrams are as follows: CG—control grid; P—plate; SG—screen grid; K—cathode; F—filament; SU—suppressor (this is tied to cathode or filament either inside or outside the tube); H—heater; D—diode (one of plates in double purpose tube).

In general, it does not matter how the filament is connected to its battery; some tube sockets have one filament prong marked plus, the other minus, but these markings need not be followed. Likewise in heater type tubes being operated from direct current polarity is immaterial.

Remember that it is only the last two numerals in a tube number which are important. Thus the 130, 230, 330, and 430 tubes are all the same as the '30 tube. The first number merely indicates who manufactured that tube.

Newer tubes are assigned three-symbol designations. The first numerals indicate the filament voltage, 1 representing filament voltages between 1 and 2 volts, 6 indicating a 6.3 volt tube, etc. The central letter merely indicates the type of tube, Z standing for a rectifying tube, etc. The last figure indicates the number of different electrodes in the tube, the filament here being considered as one electrode.

**CORRECT FILAMENT VOLTAGES**

| | | |
|---|---|---|
| 1.5 volts | A.C. | '26 |
| 2.0 " | D.C. | '30, '31, '32, '33, '34, '49 |
| 2.5 " | A.C. | '24A, '27, '35, '45, '46, '47, '55, '56, '57, '58, '82, 2A3, 2A5, 2A6 |
| 3.3 " | D.C. | '20, '22, UX'99, UV'99 |
| 5.0 " | A.C. or D.C. | '00A, '01-A, '12-A, '40, '71-A, '80, '83, 5Z3 |
| 6.3 " | A.C. or D.C. | '36, '37, '38, '39, '41, '42, '43, '75, '76, '77, '78, '85, 6A4, 6C6, 6D6 |
| 7.5 " | A.C. or D.C. | '10, '50 |

# Check Antenna and Ground

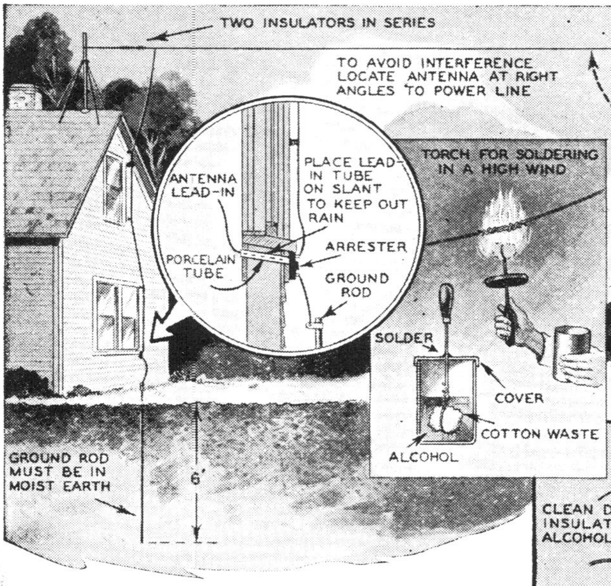


Power lines often create annoying interference in radio sets. This can be reduced by locating antenna at right angles to power wires, as shown above. Useful kink for soldering antenna or lead-in in high wind is shown in center.

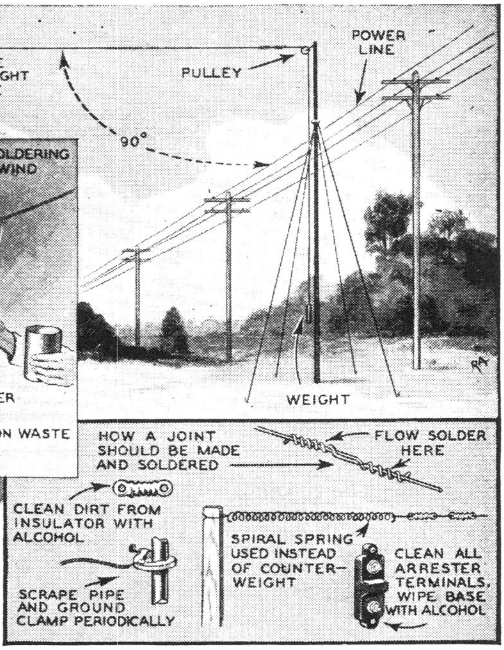


Make periodic inspection of antenna and ground system to obtain trouble-free DX reception with your radio set. Pulley and weight arrangement or spring keeps antenna tight.

WHEN the reception of stations on your radio set is accompanied by annoying noises, or your radio stops altogether, there is a good chance that the trouble is either in the antenna or the ground system.

A few simple tests which can be made without tools or test equipment will tell you where to begin looking for trouble.

**Simple Test Isolates Trouble**

Radio signals are picked out of the air by your antenna, so it is only logical you look there first. Where interfering noises are heard with radio programs, disconnect the antenna wire at the set. If the noise disappears entirely, you have located the trouble. If it stays the same, or becomes fainter in direct proportion to the reduction in radio signals when the antenna is disconnected, look for the trouble elsewhere.

In most cases where the antenna is at fault the trouble can be located by inspection. Look for loose contacts, especially at the point where the lead-in is fastened to the antenna, and at any joints which may be present in the circuit.

If antenna or lead-in is rubbing on some tree or building, there will be a leakage of radio-frequency signals to the ground; this will reduce volume.

Lead-in wires which sway in the wind may become crystallized at certain points, and eventually develop a break which makes intermittent contact and produces crashing noises in the set. Where this is suspected, go over every inch of the lead-in. Add more stand-off insulators to stop this swaying.

In many homes the radio set becomes louder whenever lights or appliances are turned on in a distant part of the house. When the lights are turned off, the volume drops back to its normal level.

This change in volume is due to the light wires themselves serving as antenna. More wires are added to this power line antenna when switches are closed, with the resultant increase in reception volume. Once a good outside aerial is installed and connected properly to the set, the change in volume will no longer be noticeable.

**Good Grounds Are Required to Get DX**

The ground is just as important as the antenna in bringing in clear DX reception. The lower the resistance between the ground lead and the ground, the greater will be the strength of the signals received.

This simple test will tell you whether you have a good ground connection. Tune in the whistle of some distant radio station, then touch the ground binding post of the

# To Get DX RECEPTION

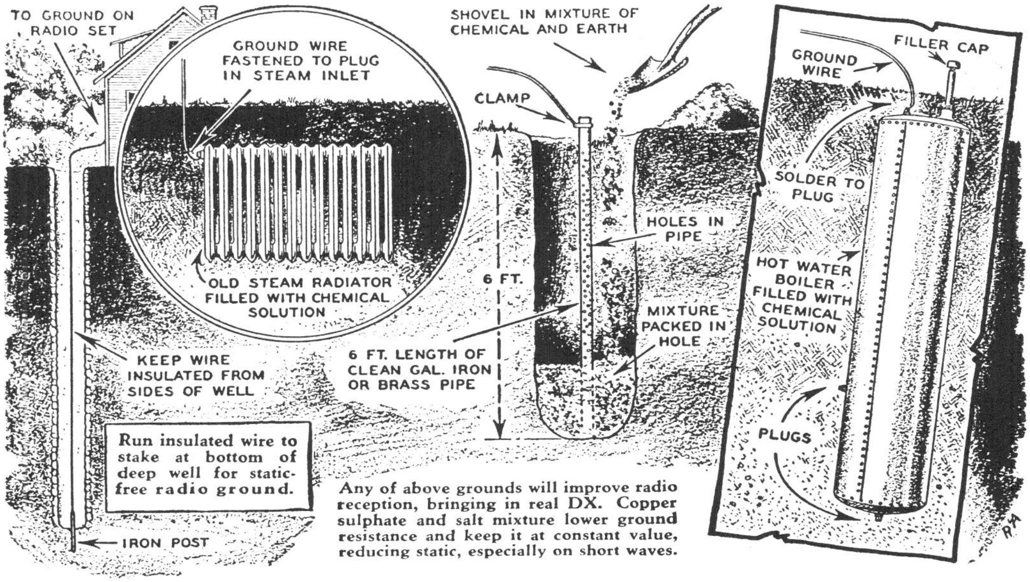


Run insulated wire to stake at bottom of deep well for static-free radio ground.

Any of above grounds will improve radio reception, bringing in real DX. Copper sulphate and salt mixture lower ground resistance and keep it at constant value, reducing static, especially on short waves.

set several times with your finger. If the ground connection is poor, you will be able to detect a decided variation in the pitch of the transmitting station's whistle.

First, of course, the ground clamp should be checked for dirty, corroded, or loose contact. If your ground happens to be in a location which is sandy or dry, reception will be poor. Pour plain or salt water around the ground pipe; if reception improves after allowing a few minutes for the water to sink in, a longer pipe should be used, or the ground moved to a damper location. The water here improves the conductivity of the earth.

Where your radio is simply grounded to a nearby steam radiator or water pipe, try the finger test mentioned above. In steam systems, packing material is sometimes used between sections of pipe; this, of course, breaks up your ground system. A length of copper wire soldered across the joint will help, if the trouble-making point can be located.

### Special Ground Reduces Static

Experiments have proved that a radio ground established far down into the earth, with heavily insulated lead-down wires, keeps radio static down to a minimum even during thunder storms.

Electro-chemical grounds are not hard to make, and are ideal in locations where the ground is too dry during the summer. Dig a six-foot deep hole in damp ground, and place in a length of copper or galvanized pipe having a number of ¼″ holes drilled through it. Now mix up 3 lbs. copper sulphate, 5 lbs. rock salt and 3 lbs. pulverized charcoal with the earth removed. Replace this mixture in the hole, tamping down firmly. Fill the pipe with water every week during the dry season.

Another good ground is made by cleaning the paint off an old steam radiator or water tank. Plug up all openings, and bury in damp ground. Fill with water containing all the previously mentioned chemicals except the charcoal. It will be necessary to replace the evaporated water occasionally, but the chemicals need not be added unless there are leakage holes in the tank.

Single wire can be used to connect magnetic speaker in garage or separate room to your radio. Ground itself serves as other lead. Make connections as shown above, placing 2 mfd. condenser in series with line to prevent accidental grounding of wire. Method applies only to sets having magnetic speakers. Both set and extra speaker must have ground.

# HUNTING for TROUBLE

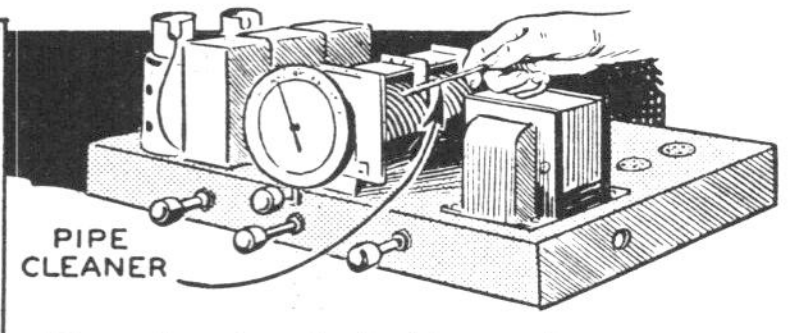


Pipe cleaners obtainable at cigar stores are handy for brushing out dust from plates of variable condensers.

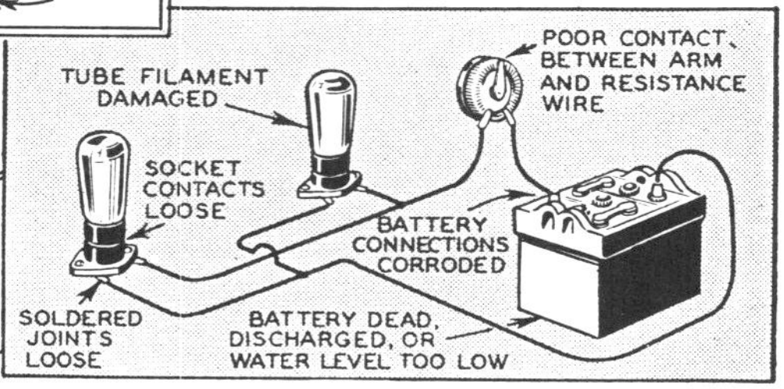


Adjustable radio set cradle shown above makes trouble-hunting easier, for set can be tilted or turned to any convenient position. Pad thumb-screws and metal side plate to prevent scratches on cabinet.

Above sketch indicates where to look for trouble in filament circuit of battery set. If speaker faces radio set, as at left, sound vibrations may react on tubes to produce annoying acoustic feedback howls. Remedy is to move speaker away, or to mount tube sockets on sponge rubber pads.

*by* L. J. MARKUS

BEFORE looking for trouble inside a radio set, be sure you have checked over the entire antenna and ground systems as outlined in the preceding article. By eliminating, step by step, places where the trouble in question cannot be, your work is greatly simplified.

Causes of trouble in the different types of radio sets—battery or electric, short wave or long wave—are in general just about the same. The type of trouble in which there is no response whatsoever in the phones or speaker is perhaps the most puzzling to the radio amateur, yet this is no more difficult to locate and repair than a case where the set is working poorly.

Systematic checking of your radio set in the manner described in this article will locate any of the more common battery set troubles, and in a great many cases clear up all-electric receiver troubles.

### Why New Receivers Fail

Where a set has been in operation for less than two weeks, look out for defective vacuum tubes, broken or incorrect battery connections, or loose speaker connections. In the case of an electric set, a burned out power transformer because of too high line voltage is a common occurrence.

Neglect in keeping the A battery properly charged is probably the reason for failure of a battery radio after two or three months' operation. B batteries should be replaced about this time, too.

After about a year of operation, test all of the tubes before looking any farther for trouble. Where breakdown is sudden, the trouble is more likely to be mechanical—either a broken tube filament, or a loose connection somewhere.

Dust on the plates of the variable condensers will reduce the volume; ordinary felt pipe cleaners are best for cleaning dust in between the plates.

Poorly soldered connections will begin to corrode in a salt atmosphere such as exists near the ocean. Resoldering of all connections in the set is the cure here.

Where cheap grades of insulating materials have been used, radio frequency currents will be short-circuited by the insulation itself in damp weather, reducing the strength of signals. A slow drying of the set in a warm oven will stop this trouble, at least for a time.

Worn mechanical parts should be looked for when sets which have been working good for several years suddenly fail. Ver-

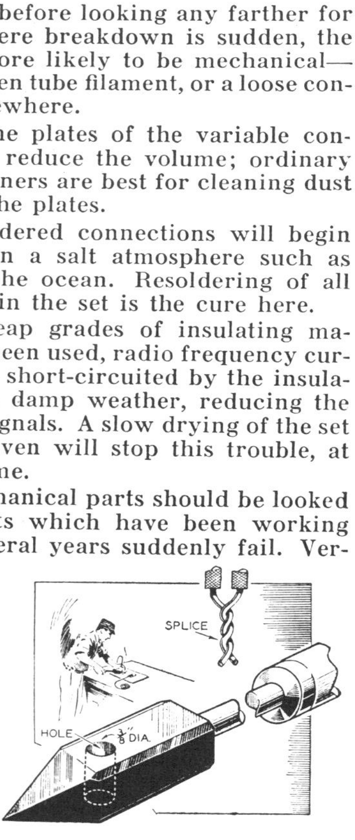


To speed up soldering of wire splices in radio repair work, drill 3/8" diameter hole in soldering iron, but not all the way through. Partly fill hole with solder, then dip splices into solder, after first applying soldering paste to the wires.

# In Your RADIO RECEIVER

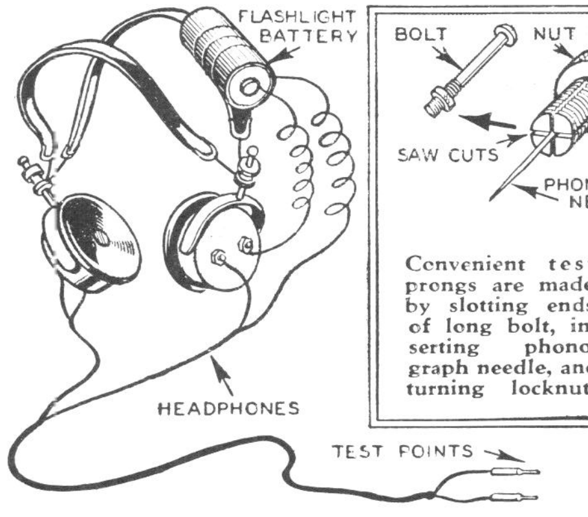


If flashlight cell is strapped to band of headphones and connected in series with one of phone leads as shown, phone tips are ready for use as test points.

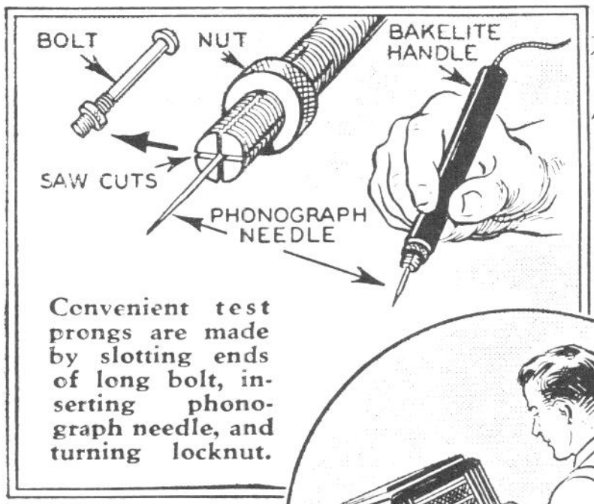


Convenient test prongs are made by slotting ends of long bolt, inserting phonograph needle, and turning locknut.

Where loose connection is suspected as cause of trouble, yet cannot be located by ordinary methods, resolder every connection in set by touching hot iron to each joint for few seconds. Loose connection is indicated if crackle is heard in phones or speaker when set is turned on and jarred.

nier dials sometimes get loose, producing noises like those due to loose connections. Vacuum tubes are, of course, to be checked here, as well as batteries.

The filament circuit is the first and easiest thing to check when trouble develops. An open circuit somewhere is indicated if the tubes flicker or fail to light up, once the batteries have been checked for voltage or for loose or corroded connections.

### Inspect All Tube Sockets

Spring contacts in the tube sockets should be sanded lightly, and the tube prongs cleaned. Bend the contacts together to make them press harder against the tube prongs, where loose connections are suspected.

If there is a rheostat in the filament circuit, increase the tension of the contact arm, and sandpaper the windings lightly.

Good reception cannot be expected if the batteries are weak. Modern tubes are designed to be operated at a specific filament voltage; while they will give satisfactory operation at values considerably below the correct one, the life of the tube will be seriously impaired.

A short circuit is a common cause of trouble where there is no response whatsoever in the phones or speaker. Rotor and stator of a variable condenser may be touching, particles of solder or wire may have fallen between the plates, or a loose bolt may be lodged across two wires in some part of the set. Look for these troubles if a noise is heard when the set is jarred.

Too small a condenser across the headphones, or too small a grid condenser will give faint signals accompanied by a scratching or fuzzy noise. A small fixed condenser connected in parallel with the one already in the set will clear up this trouble.

Weak signals or none at all, accompanied by a humming sound in the phones, may usually be traced to an open circuit in the grid circuit. Look for a crack in the grid resistor, or a broken wire somewhere near the tube grid.

### Test B Batteries With Headphones

Defective B batteries may be located by connecting headphones across the battery. Only one click will be heard when contact is made if the battery is good. Defective or run-down batteries will give a series of scratchy noises.

Where signals break off with a popping sound, disappear entirely for a few seconds, then come back and disappear repeatedly, there is either an open wire in the grid circuit or too high a value of grid leak resistance. A defective C battery will also give this effect.

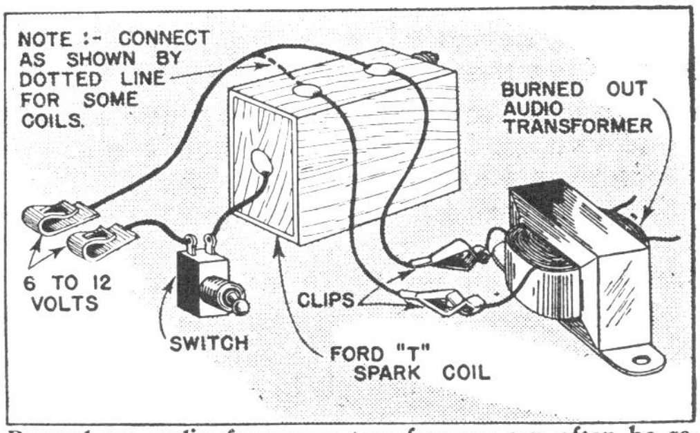


Burned-out audio frequency transformers can often be repaired by connecting across high voltage secondary of model T Ford spark coil. Spark will jump across broken wires inside transformer, perhaps melting enough of wire to bridge gap. Chokes can oftentimes be repaired in same way.

# WHAT CAN I EXPECT

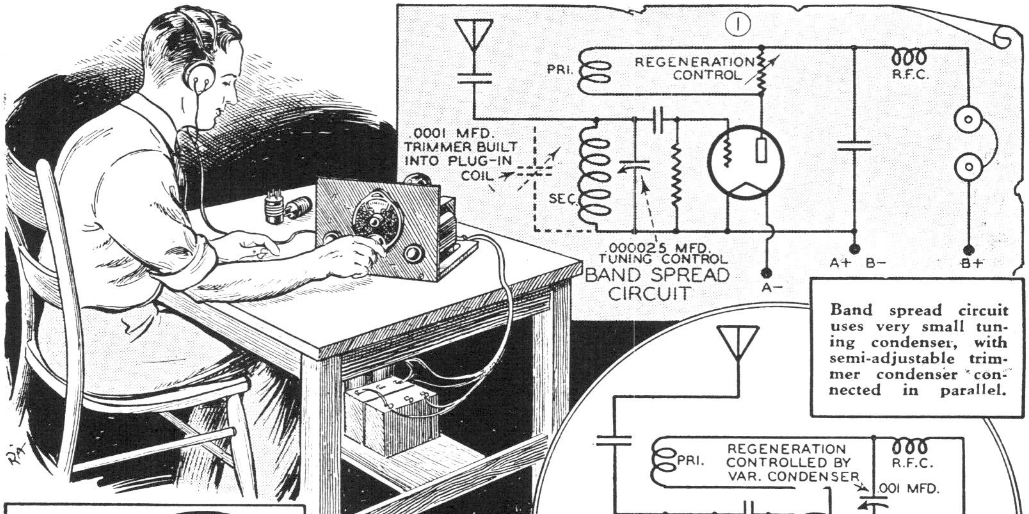


Band spread circuit uses very small tuning condenser, with semi-adjustable trimmer condenser connected in parallel.

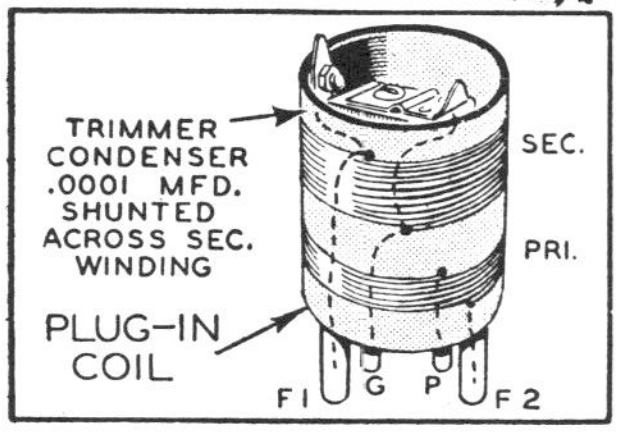


To obtain band spread with each plug-in coil, mount trimmer condenser inside as shown at left. Adjust with fiber screwdriver while midget tuning condenser is set at minimum.

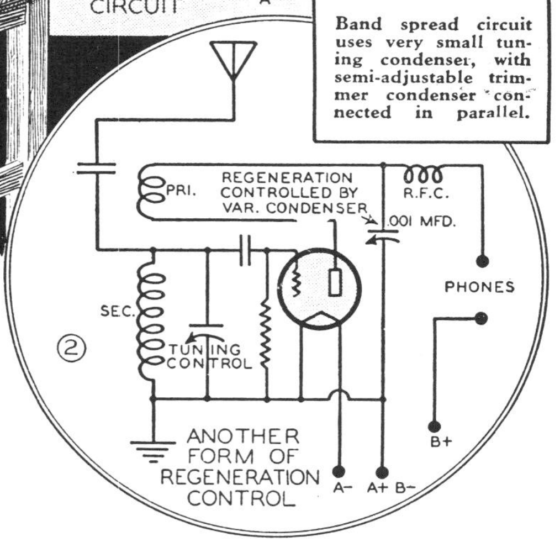


In above circuit smooth regeneration control is obtained by variable condenser in plate circuit. Resistor control of regeneration is shown in top circuit. Circuit constants have not been given, since they can be found in other articles.

IN SHORT WAVE tuning there are many things to take into consideration before judgment can be pronounced on a particular short wave radio set. The short waves being so elusive, many experimenters and listeners give up after a few trials. A few simple precautions make it possible to hear those stations on the other side of the world that are so much discussed, but never received.

Simple one and two tube receivers can and will bring in foreign broadcasts. While too much cannot be expected of a simple set, it is, nevertheless, a stepping stone to something more elaborate.

The simplest type of short wave circuit is shown in Figure 1, this being nothing more than a straight regenerative receiver. With this circuit it is possible to add both audio and radio frequency stages to finally obtain a powerful short wave receiver.

In the process of adding audio frequency stages to a detector there are two methods of coupling from which to select. The first is resistance coupling, and the second transformer coupling. If fine tone is desired you should choose resistance coupling, but if you want volume and distance select the transformer coupled type. The two types are shown in circuits 5 and 7; a type '26 tube or any other amplifier tube may be used, depending upon the filament voltage which is available.

Owners of good battery operated broadcast receivers may use the audio frequency stages of their radio, by using the 1-tube regenerative detector as an adapter. Since most sets have at least two stages of audio frequency the simple 1-tube battery set, when coupled to the broadcast set, will actually be a 3-tube short wave receiver. Other adapter circuits, these having all circuit constants, are shown on pages 22 and 74 of this book.

In order to get the most out of your short wave set, a stage of tuned radio frequency is suggested. This is placed between the antenna and the detector stage to amplify r.f. signals before detection, giving a great increase in volume. This r.f. stage is shown in circuit 3, and is very similar to the de-

# of My Short Wave Radio?

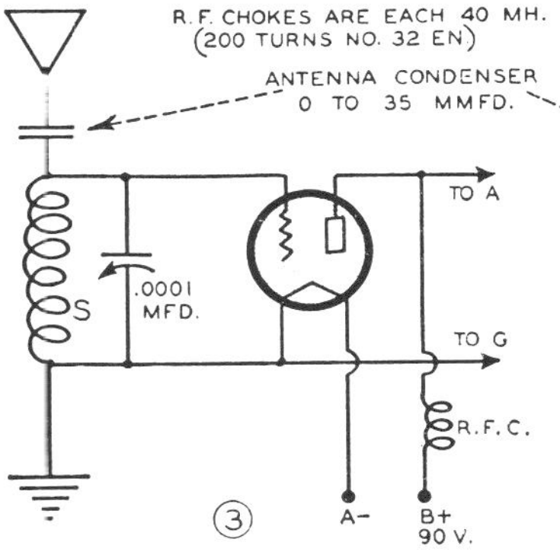

R. F. STAGE FOR GREATER POWER

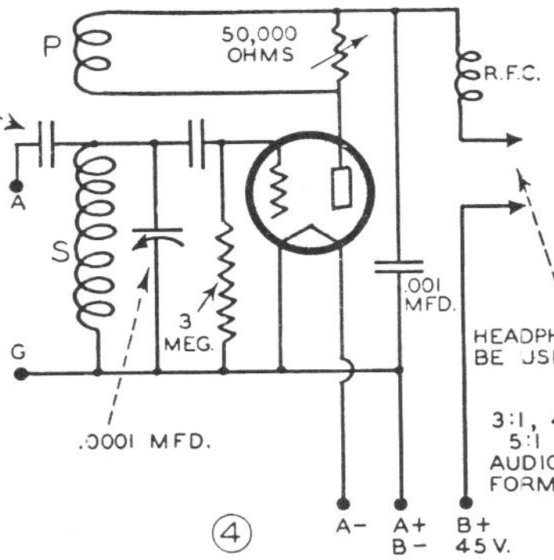

DET. STAGE MAY BE USED AS A 1-TUBE SET

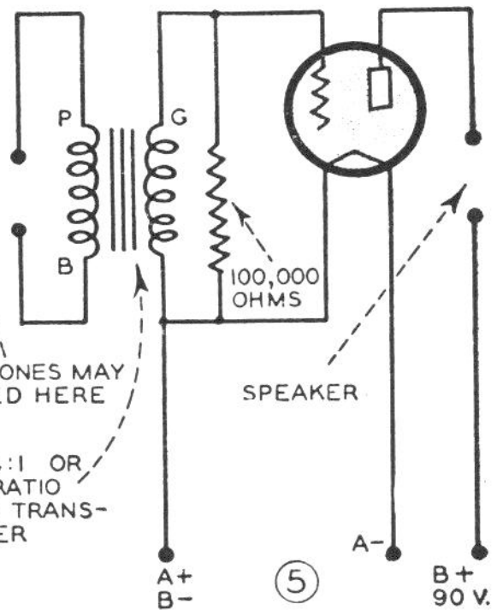

TRANSFORMER COUPLED A. F. STAGE

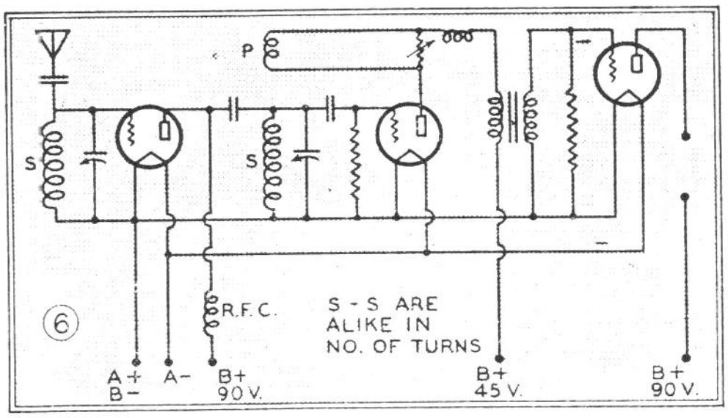


Radio frequency, detector, and audio frequency stages are shown separately in top diagrams, and combined into one powerful 3-tube set at left. Resistance coupled audio frequency stage, shown at right, may be substituted for circuit 5 where better tone with less volume is desired. Transformer must be connected exactly as shown to get results. Signals are amplified many times in passing through transformer.

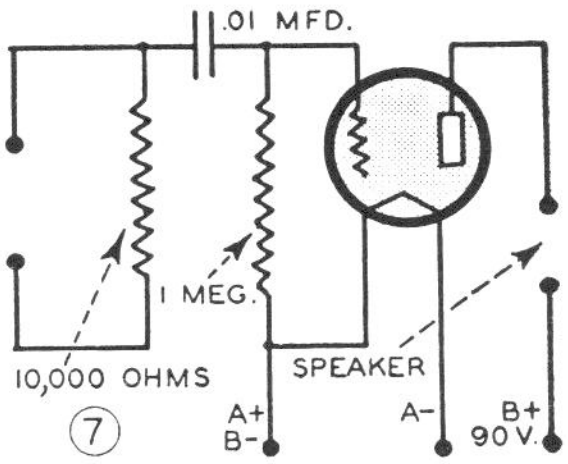

RESISTANCE COUPLED A. F. STAGE

tector stage, except that the plug-in coil has only one winding. This winding matches the secondary winding on the regular plug-in coil, but does not use a plate winding. It is necessary to change two coils each time the wave band is changed, when using this set.

Circuit diagrams 3, 4 and 5 show how each of the above-mentioned units appear separately. In circuit 6 they are all connected together to work as one unit. Either the a.f. or r.f. stages, or both, may be added to your own one-tube set.

Results in short wave work depend a great deal upon location. In places bordering on our eastern or western seaboard, reception of trans-oceanic broadcasts is possible with a single 1-tube regenerative detector. With a stage of transformer coupled a.f. added, stations 12,000 miles away have been listened to with ease.

The antenna is very important in one sense, and in another it is not. The writer has received European broadcasts on an antenna only one yard long on several occasions—and the set had only one tube. Of course, atmospheric conditions must be favorable, and the set carefully tuned.

For the newcomer a good antenna is suggested. This should be erected as high as possible and as far away from power lines as is practical. For the receivers described here the antenna is simply a length of No. 14 or No. 12 enameled copper wire about 100 feet long. Solder the lead-in to one end, and drop it directly down to the receiver. For sets like those described here doublet antennas are not recommended.

Only by practice can you gain skill in tuning a regenerative short wave receiver. After experimenting with your set for a week you will be able to clear up the faint whistles and bring in those foreign broadcasts.

A short wave log book is very necessary, since no foreign stations can be picked up if they are not on the air.

With ordinary tuning controls all of the stations will come in within two or three dial divisions. To spread these stations out over your entire dial, use an extremely small tuning condenser, and use a trimmer condenser connected across the coil secondary to bring the capacity up near the point where stations are heard with that particular coil. Oftentimes these trimmers are built into the plug-in coils, each being adjusted once to match with the tuning condenser.

# U. S. Police Station Log

**This list of all police transmitters operating in the United States will help you in identifying the different calls. With patience you should be able to pick up almost all of these stations with a good short wave receiver, adapter, or converter.**

| LOCATION and CALL | Kilocycles |
|---|---|
| Aberdeen, Wash., KGZV | 2414 |
| Akron, Ohio, WPDO | 2458 |
| Albany, N. Y., WPGH | 2414 |
| Albuquerque, N. M., KGZX | 2414 |
| Arlington, Mass., WPED | 1712 |
| Atlanta, Ga., WPDY | 2414 |
| Auburn, N. Y., WPDN | 2382 |
| Austin, Texas, KGHU | 2382 |
| Bakersfield, Cal., KGPS | 2414 |
| Baltimore, Md., WPFH | 2414 |
| Bay City, Mich., WPCA | 2466 |
| Beaumont, Texas, KGPJ | 1712 |
| Berkeley, Cal., KSW | 1658 |
| Binghamton, N. Y., WPGL | 2442 |
| Birmingham, Ala., WPFM | 2382 |
| Bridgeport, Conn., WPFW | 2466 |
| Buffalo N. Y., WMJ | 2422 |
| Cedar Rapids, Iowa, KGOZ | 2466 |
| Chanute, Kans., KGZF | 2450 |
| Charlotte, N. C., WPVD | 2458 |
| Chicago, Ill., WPDB | 1712 |
| Chicago, Ill., WPDC | 1712 |
| Chicago, Ill., WPDD | 1712 |
| Cincinnati, Ohio, WKDU | 1706 |
| Clarksburg, W. Va., WPFP | 2490 |
| Cleveland, Ohio, WRBH | 2458 |
| Coffeyville, Kans., KGZP | 2450 |
| Columbus, Ga., WPFI | 2414 |
| Columbus, Ohio, WPGQ | 1682 |
| Cranston, R. I., WPGK | 2466 |
| Dallas, Texas, KVP | 1712 |
| Davenport, Iowa, KGPN | 2466 |
| Dayton, Ohio, WPDM | 2430 |
| Denver, Colo., KGPX | 2442 |
| Belle Isle (Det.) Mich., WCK | 2414 |
| Detroit, Mich., WPDX | 2414 |
| Des Moines, Iowa, KGZG | 2466 |
| Washington, D. C., WPDW | 2422 |
| East Lansing, Mich., WRDS | 1666 |
| E. Providence, R. I., WPEI | 1712 |
| El Paso, Texas, KGZM | 2414 |
| Flint, Mich., WPDF | 2466 |
| Fort Wayne, Ind., WPDZ | 2490 |
| Columbus, Ohio, WPDI | 2430 |
| Fresno, Cal., KGZA | 2414 |
| Grand Rapids, Mich., WPEB | 2442 |
| Grosse Pointe, Mich., WRDR | 2414 |
| Hackensack, N. J., WPFK | 2430 |
| Highland Park, Mich., WMO | 2414 |
| Honolulu, T. H., KGPQ | 2450 |
| Huntington, N. Y., WPGO | 2490 |
| Indianapolis, Ind., WMDZ | 2442 |
| Johnson City, Tenn., WPGZ | 2474 |
| Jacksonville, Fla., WPFG | 2442 |
| Kansas City, Mo., KGPE | 2422 |
| Klamath Falls, Ore., KGZH | 2382 |
| Knoxville, Tenn., WPFO | 2474 |
| Kokomo, Ind., WFDT | 2490 |
| Lansing, Mich., WPDL | 2442 |
| Las Vegas, Nev., KGHG | 2474 |
| Lexington, Ky., WPET | 1706 |
| Lincoln, Neb., KGZU | 2490 |
| Little Rock, Ark., KGHZ | 2406 |
| Los Angeles, Cal., KGPL | 1712 |
| Louisville, Ky., WPDE | 2442 |

| LOCATION and CALL | Kilocycles |
|---|---|
| Lubbock, Texas, KGZW | 2458 |
| Memphis, Tenn., WPEC | 2466 |
| Miami, Fla., WPFZ | 2442 |
| Milwaukee, Wis., WPDK | 2450 |
| Minneapolis, Minn., KGPB | 2430 |
| Muncie, Ind., WPGP | 2442 |
| Muskegon, Mich., WPFC | 2442 |
| Mineola, N. Y., WPGS | 2490 |
| Fairhaven, Mass., WPFN | 1712 |
| New Orleans, La., WPEK | 2430 |
| Newton, Mass., WPFA | 1712 |
| Brooklyn, N. Y., WPEE | 2450 |
| New York, N. Y., WPEF | 2450 |
| New York, N. Y., WPEG | 2450 |
| Oklahoma City, Okla., KGPH | 2450 |
| Omaha, Neb., KGPI | 2466 |
| Palm Beach, Fla., WPFX | 2442 |
| Palo Alto, Cal., KGHK | 1682 |
| Pasadena, Cal., KGJX | 1712 |
| Pawtucket, R. I., WPFV | 2466 |
| Philadelphia, Pa., WPDF | 2474 |
| Phoenix, Ariz., KGZJ | 2430 |
| Pittsburgh, Pa., WPDU | 1712 |
| Port Huron, Mich., WPGB | 2466 |
| Portland, Me., WPFU | 2422 |
| Providence, R. I., WPGF | 1712 |
| Portland, Ore., KGPF | 2442 |
| Portsmouth, Ohio, WPGI | 2430 |
| Richmond, Ind., WPDH | 2442 |
| Reading, Pa., WPFE | 2442 |
| Rochester, N. Y., WPDR | 2422 |
| Rockford, Ill., WPGD | 2458 |
| Saginaw, Mich., WPES | 2442 |
| St. Louis, Mo., KGPC | 1706 |
| St. Paul, Minn., WPDS | 2430 |
| Salem, Ore., WGZR | 2442 |
| Salt Lake City, Utah, KGPW | 2406 |
| Santa Barbara, Cal., KGZO | 2414 |
| Santa Cruz, Cal., KGZT | 1674 |
| San Antonio, Texas, KGZE | 2482 |
| San Bernardino, Cal., KGZY | 1712 |
| San Diego, Cal., KGZD | 2490 |
| San Francisco, Cal., KGPD | 2466 |
| San Jose, Cal., KGPM | 1674 |
| Santa Fe, N. M., KGPF | 2414 |
| Seattle, Wash., KGPA | 1414 |
| Sioux City, Iowa, KGPK | 2466 |
| Somerville, Mass., WPEH | 1712 |
| South Bend, Ind., WPGN | 2490 |
| Spokane, Wash., KGHS | 2414 |
| Swarthmore, Pa., WPFQ | 2474 |
| Syracuse, N. Y., WPEA | 2382 |
| Tacoma, Wash., KGZN | 2414 |
| Toledo, Ohio, WRDQ | 2474 |
| Topeka, Kans., KGZC | 2422 |
| Tulare, Cal., WPDA | 2414 |
| Tulsa, Okla., KGPO | 2450 |
| Utica, N. Y., WPGJ | 2414 |
| Vallejo, Cal., KGPG | 2422 |
| Waco, Texas, KGZQ | 1712 |
| Wichita Falls, Texas, KGZI | 2458 |
| Wichita, Kans., KGPZ | 2450 |
| Woonsocket, R. I., WPEM | 2466 |
| Worcester, Mass., WPGX | 2466 |
| Youngstown, Ohio, WPDG | 2458 |

# Locating Troubles in SHORT WAVE SETS

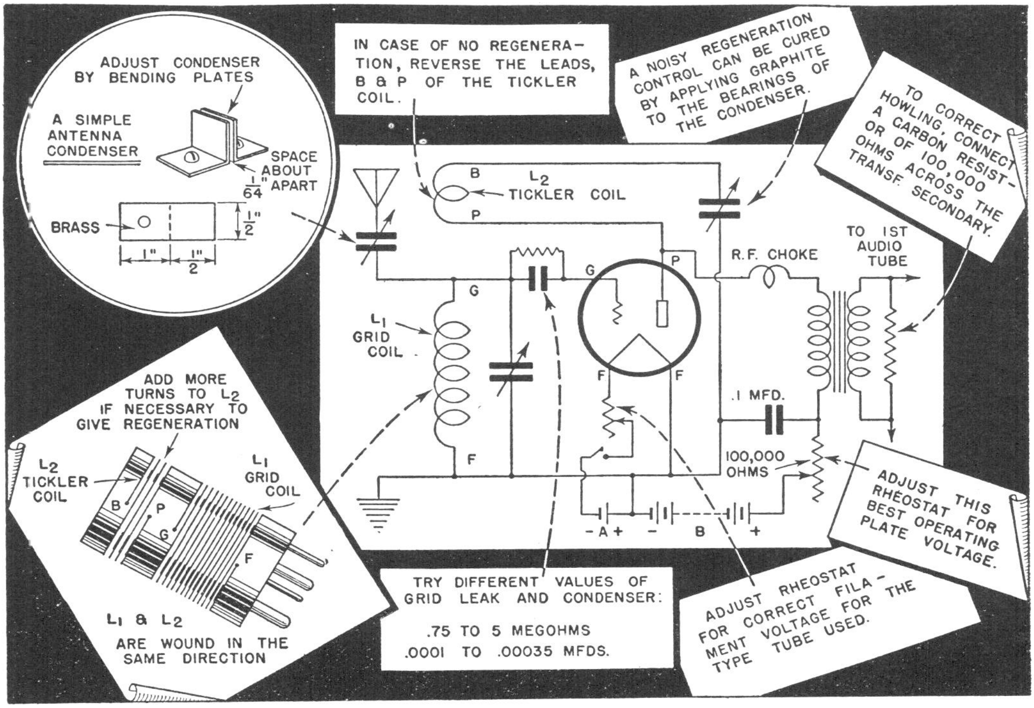


Common troubles of a regenerative short wave receiver circuit are labeled above, with their cures. Bookplate antenna condenser, construction of which is shown, is adjusted once if only one coil is used in set, then left fixed, since it merely tunes antenna to set. This condenser should be readjusted each time plug-in coils are changed, to get maximum volume and distance. Importance of proper coil connections cannot be stressed too much. Windings should be same direction.

THIS article will point out some of the more common mistakes made by radio fans in building short wave sets or converters. Many are the builders who thought they had followed diagrams exactly, yet obtained a set which gave very disappointing results.

First of all, if you are building a short wave receiver, a good aerial is necessary. A small antenna condenser, variable over a range of 5 to 35 mmfd., will materially increase the selectivity of the set. The small capacity is needed for the short waves, and the larger capacity, with the plates close together, for the broadcast band.

If you are building a converter—this being simply a detector stage which feeds into the audio stages of your own radio set—do not expect good results if the set itself cannot get good distance. Midget receivers are seldom satisfactory for short wave work.

It will be impossible to get any regeneration if the tuning coils are wound or connected incorrectly. Be doubly sure that in winding your coils you make both tickler and grid coils turn in like directions.

If the set will not regenerate, reverse the connections to the plate winding of the plug-in coil. If this does not help, increase the number of turns on the plate (tickler) winding until the set does regenerate.

If just the right plate voltage is used in a regenerative detector, the set will be far more sensitive and easier to operate. For most tubes 45 volts is used; this may be varied by connecting a 100,000 ohm rheostat in the plate lead to the detector tube, and placing a .1 mfd. condenser from the rheostat to ground as shown in the above circuit. A filament rheostat will also help to control regeneration, if the detector tube is of the filament type.

A carbon resistance of about 100,000 ohms, connected across the secondary of the first audio frequency transformer, will correct annoying howls due to feedback.

If the set works poorly, try different values for the grid resistances and capacities, ranging from .75 to 5 megohms and from .0001 to .00035 mfd.

If 2-volt tubes are used, never go over this voltage on the filament—from 1.5 to 1.75 volts will be better.

# SIX-TUBE Auto RADIO

Tuning and volume control dials are mounted on end of can. Extension coupling is used on condenser rotor to permit chassis to be slid out of can. Dynamic speaker is separate.

**Save as much as twenty-five dollars by building your own auto radio. Small, simple, and inexpensive though this 6-tube receiver is, it will bring in stations thousands of miles away with loud speaker volume.**

WHEN building an auto receiver several very important factors must be considered or else the receiver will be too large, not sensitive or too costly to construct. To be certain that none of these difficulties will arise, a receiver has been designed that is very sensitive, tiny, simple to construct and, most important, very inexpensive.

The average person does not want to work with complicated circuits which are even too difficult for an experienced radio engineer. In this receiver the circuit has been made just as clean-cut as possible with all unnecessary by-pass condensers and resistors left out. Why complicate a circuit with things that do little or no good?

This auto radio employs six inexpensive automotive tubes. These include two –78 R. F. amplifiers, a –77 as a grid bias detector, a –37 as audio amplifier and two –38 power amplifiers in push-pull. This arrangement gives plenty of volume for auto and even more when used as a farm set with an antenna.

Most of the materials required for constructing this receiver can be purchased from any mail order radio supply house. The chassis and shield can can be constructed by the builder or by a tinsmith. In some cases it may be possible to secure a finished chassis from the radio supply house.

In purchasing the necessary coils be certain that they match the tuning condensers. If you are very ambitious you can wind your own coils, but it is suggested that you buy them ready made since the set of three matched coils in aluminum shield cans costs less than $1.50.

For those wishing to make their own the following is given: Secondary windings of R. F. coils—127 turns of No. 36 enameled copper wire wound on a bakelite or even cardboard form (mailing tube) 1″ in diameter. The primary coil is wound over the secondary winding, separated by insulating cloth (commonly called "Empire" cloth) or "fish" paper. In the case of the antenna coil the primary winding is a manufactured type choke coil mounted on the inside of the form. A R. F. choke of 40 M.H. will be satisfactory. If at all possible it is best to use ready made coils as good results cannot be assured with homemade affairs. The coils are mounted in copper, aluminum or zinc cans size 2″ in diameter and about 2½″ high.

Photograph of completed chassis is shown above. R. F. and detector tubes are shielded in metal "stalls."

The dynamic speaker is a separate unit, since the average radio experimenter is not able to secure the midget speakers used in commercial one piece auto radios.

The power unit has also been left out of the chassis, permitting the builder to use either B batteries or a B power unit operating from the car battery, as he wishes. Plans for building your own B power unit appear on the opposite page.

The chassis is made according to the template shown. It is nothing more than a sheet of heavy galvanized iron or, better,

# BRINGS in DX

By THOMAS A. BLANCHARD

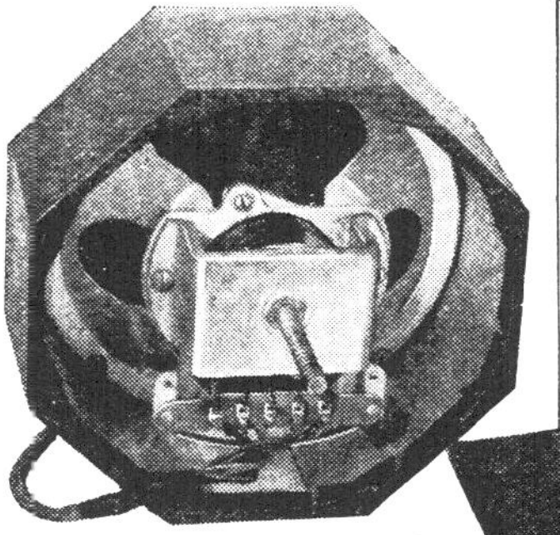

Make connections to terminal panel of 6" auto radio dynamic speaker according to diagram on first page. Right: Condenser shaft is cut off so chassis will slide out of can. ¼"x¼" shaft extender is used to connect shaft to tuning dial.

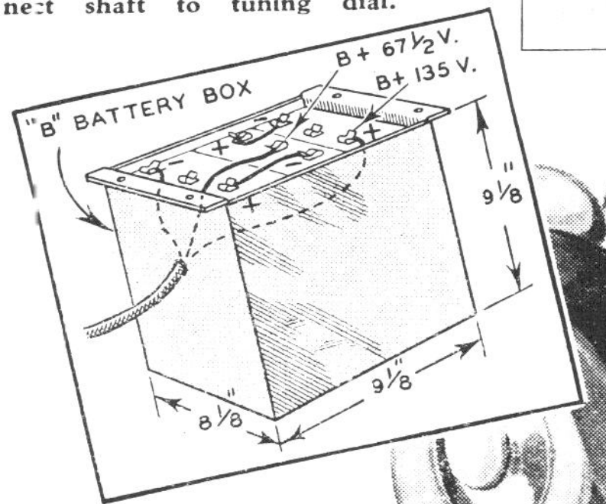


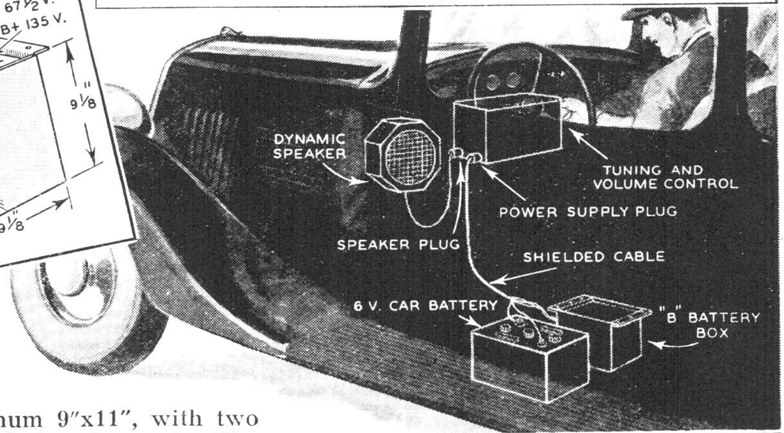


If B batteries are used, mount in metal battery box. Mount set and speaker in most convenient position in auto.

No. 16 gauge aluminum 9"x11", with two end sections 6"x2½" for the ends. The necessary socket holes are cut out with a hole cutter, and the chassis then cut out, bent to shape, and riveted together.

The chassis completed, construction on the actual set can be started. By taking it easy and following the pictorial diagram there will be no difficulties. First, mount the sockets. These are the wafer type which take up very little space and are bolted to the chassis top.

Next, install the three coils, being careful to mount the antenna coil in the corner. The coils are mounted directly underneath the space that will be occupied by the three gang tuning condenser. The only other item that is mounted underneath the chassis is the push-pull transformer. It is suggested that this transformer be left out until all wiring of sockets is completed so that there will be no difficulty in wiring in the sockets.

The tube shields and condensers are left

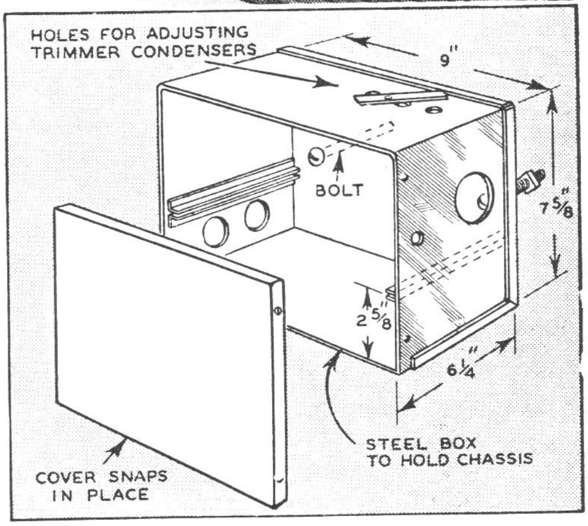


Form can or container out of heavy sheet metal. If possible have tinsmith bend metal in bar fold to secure neat corners. Rivet angle brackets in place to hold the chassis rigidly. Punch projections on cover, corresponding dents in box.

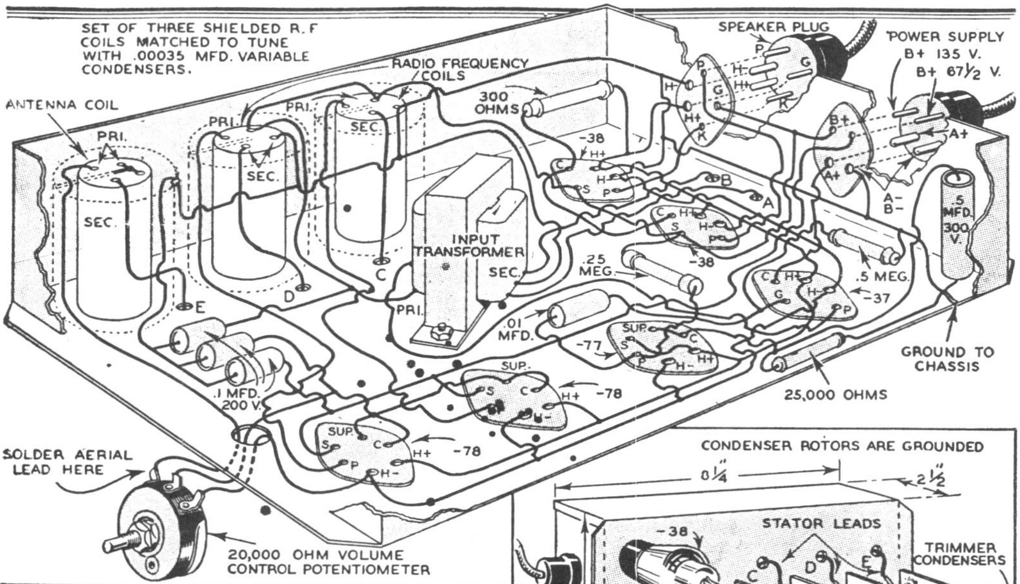


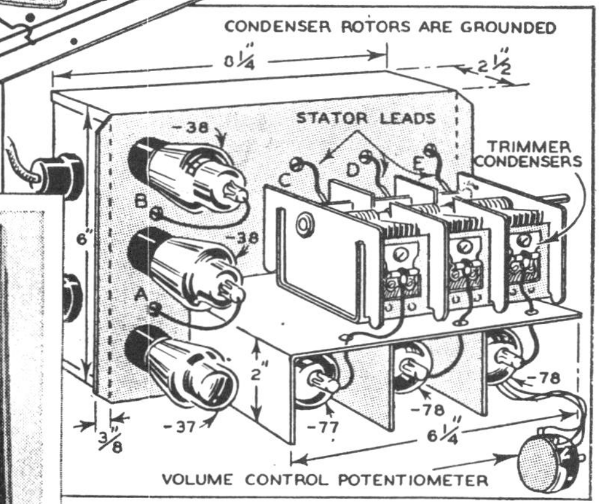


Pictorial sketches of both bottom and top of chassis show exactly how parts are connected. Follow these if you are not acquainted with radio symbols used in circuit diagram. Use insulated wire. Filament switch (not shown) mounts on back of the potentiometer, and is controlled by the same knob.

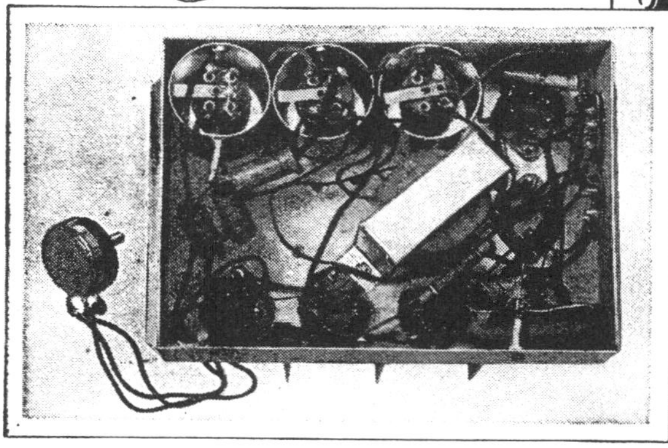

**Photo of bottom of chassis, showing arrangement of parts. Make all connections as short as possible, and solder neatly.**

off the top of the chassis until all wiring has been completed. It is much easier to wire up the set in this manner, as the chassis may be laid on the workbench bottom up and soldering in of resistors and condensers done with ease.

For ease in construction it is suggested that the fixed condensers as well as resistors be of the "pig tail" type, eliminating unnecessary wiring.

### Connect Antenna to Potentiometer

The antenna lead is connected to one side of the potentiometer arm or directly to the top of the antenna coil primary. No terminal is used here—just a soldered lead.

The two sockets on the side of the receiver chassis serve as terminals for the necessary speaker and power supply connections. Of course these schemes can be ignored and cables run directly from set to power source. The plug and socket method of making connections permits the chassis to be removed from the case at any time, without disconnecting any wires.

Now that all under wiring is completed the final work on the top of the receiver may be tackled. As the rotors of all three condenser sections are grounded to the chassis connections are only made to the fixed plates.

### Adjusting Trimmer Condensers

The three condenser gang is accurately matched at the factory, but the slightest bump or jar may destroy this balance. If a

**BLUE LINE PRINTS** for this auto radio are available at 50c postpaid. These prints will be of great help in building the set, for all diagrams are enlarged many times. A complete list of the parts needed, and instructions for making your own B power supply have been added on the blueprint. Send your order to Blueprint Dept., Modern Mechanix Publishing Co., 529 S. 7th St., Minneapolis, Minn.

# *Circuit Diagram Shows Remarkable Simplicity for Six-Tube Receiver*

gang condenser having trimmer condensers mounted integral is used, as shown in the sketches and photographs, the tuning condensers can be rematched after the set is assembled. Adjust each with a screwdriver after the set is working until reception is loudest and clearest.

Short leads are run from the trimmer condensers to their respective tube grids. Remember that in wiring in screen grid tubes the grid lead goes to the cap on top of the tube, and the screen grid lead to the grid prong of the socket.

It is very important that the radio frequency and detector tubes be shielded from each other. To do this a very simple shield arrangement is used. Strips of metal slightly higher than the tubes are riveted or bolted together to house each tube in what might be called its own "stall."

The power required for this receiver is but 135 volts. If the usual 45 volt B cells are used the 67½ volt screen grid supply is taken off from the 22½ volt clip on the second B battery in the circuit. A total of three 45 volt B batteries connected in series gives the required 135 volts.

If an auto B power unit is used a center-tapped voltage divider will provide the correct screen grid voltage. Use shielded cable for all battery leads. The speaker used is an automotive type dynamic requiring only six volts on the field coil. Connect the speaker leads to the speaker plug as shown in the pictorial diagram.

### Chassis Slides Into Cans

After the receiver is completed a can for containing it must be constructed or purchased. The size of this container is 9"x-7⅝"x6¼". The front of the can opens to allow removal of the set. Two small angle strips are bolted to the container just high enough to allow the chassis to slide in.

The can must be mounted in a convenient location inside the car. The universal place for this is just under the dash. Two holes in the back of the can are drilled for this purpose. It is only necessary to drill two holes in the wood or steel bulkhead, and bolt the can in place. In this location it is very simple to get the necessary current. The chassis being automatically grounded it is only necessary to connect on to the "hot" side of the auto's ammeter to get the current for the tube filaments, dynamic speaker field and auto B Power Unit if one is used.

The last item to be dealt with is the tuning control. Here there is a choice of one of two methods. The usual scheme is a remote control device mounted on the steering post of the auto. The remote control equipment is expensive, however, and will in no way improve the operation of the set. A simpler method, now being adopted on many new makes of auto radios, will be followed.

### Tuning Dial Mounts on Set

Direct tuning with a regular dial mounted on the side of the can is used. Tuning will be easier if the set is clamped to the side of the steering post instead of under the dash. A combination volume control and switch is mounted next to the tuning dial.

The antenna may be a length of copper fly screen mounted under the roof of the car IF it is free from screen of the "chicken wire" type used by some auto manufacturers for supporting the roof covering of the car.

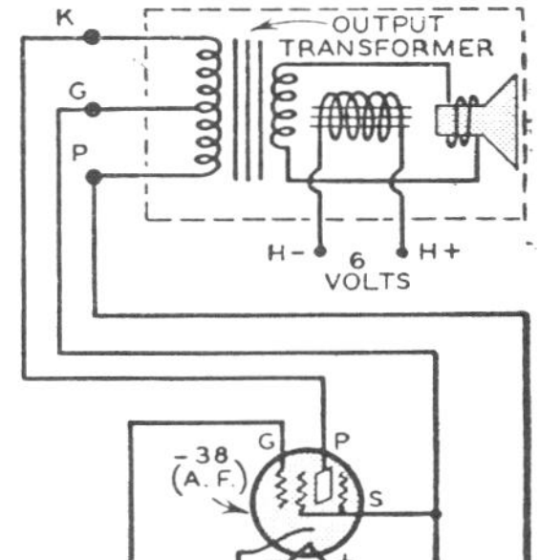


Below: Circuit diagram for 6-tube auto radio. Switch shown in A plus lead snaps onto cover of volume control. Suppressor grids of -77 and -78 tubes are connected to respective cathodes. "A-B-" is grounded throughout.

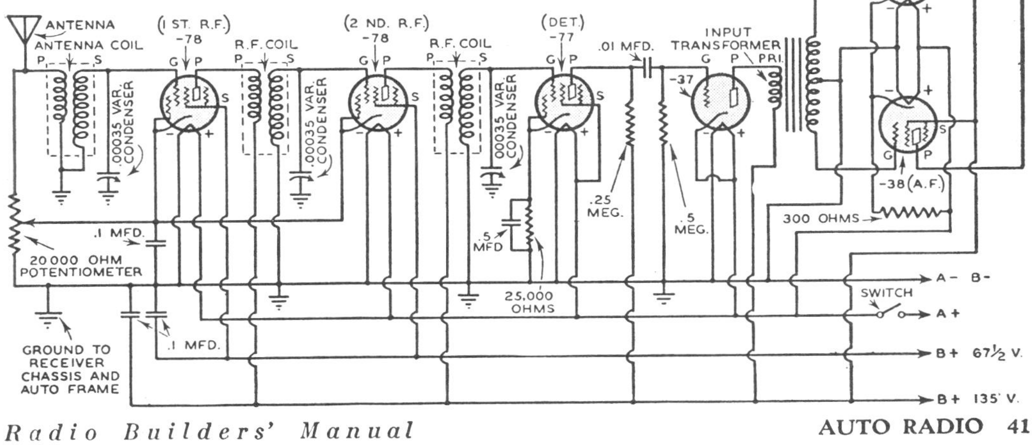

# See, Feel Radio Music With These Simple Magic Stunts

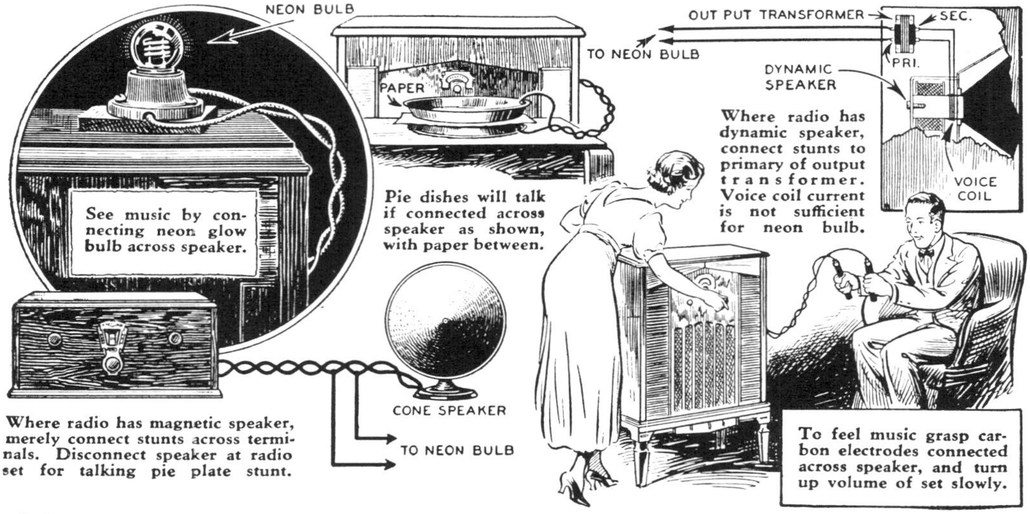


See music by connecting neon glow bulb across speaker.

Pie dishes will talk if connected across speaker as shown, with paper between.

Where radio has dynamic speaker, connect stunts to primary of output transformer. Voice coil current is not sufficient for neon bulb.

Where radio has magnetic speaker, merely connect stunts across terminals. Disconnect speaker at radio set for talking pie plate stunt.

To feel music grasp carbon electrodes connected across speaker, and turn up volume of set slowly.

A RADIO set is thought of as a single form of entertainment, but actually it is a box of electrical magic capable of performing many unusual tricks.

It is easy to fix up your radio so your friends can see music and feel it, too.

To see music get a half watt neon glow bulb at an electrical supply house. Though this bulb is small, it will screw into any light socket. Connect up the socket to the loudspeaker terminals of your radio. The bulb will glow dim on the low notes and bright when the tones are high, the brilliancy changing continually with the music.

On modern receivers connect the neon bulb ahead of the output transformer, and not to the voice coil leads.

Music can be felt as well as heard. Secure two carbon rods from old dry cells and connect one to each of the speaker leads. Moisten your hands and grip the rods. Have someone turn up the radio volume control slowly. As soon as the music is heard it will be felt. This device is a real shocking machine, so be a bit wary about gripping the handles when the volume is on full blast.

To make pie plates talk, disconnect the regular speaker, connecting the leads to the pie plates instead. Set one dish on a flat surface bottom up, place a piece of tissue paper on it, and set the other dish on next. Turn on the volume control and the dishes will talk!

---

# D.P.D.T. Switch for Electrical Experimenters Made From Spool, Screws

MANY is the time the young electrical experimenter wishes for a D.P.D.T. switch by which a source of electrical energy can be switched directly from one piece of apparatus to another. Here is just the thing for the job.

First consult the drawing shown at the right. The baseboard is four inches square, and the screws are set in a circle 3 inches square. File out the slots in the bolt heads so the heavy wire provided will lie nicely in them.

The switch is made from a spool sawed in half. Through the smaller position is run two stiff copper wires and a small brad. This latter part takes a rubber band which gives the switch elasticity. Wire the switch up according to your needs. A hookup for switching the power from one instrument to another is shown in the drawing at the right.

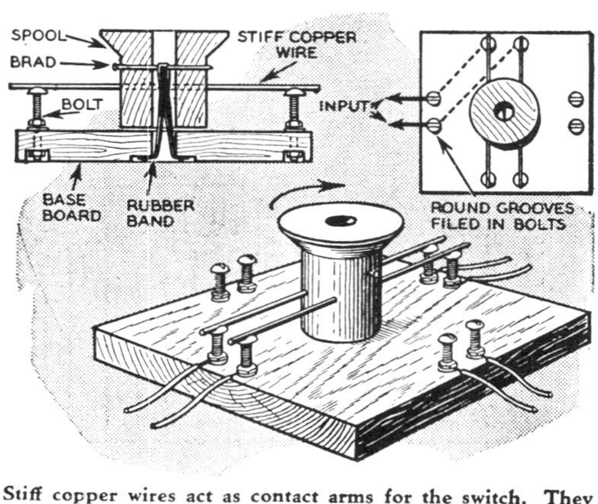


Stiff copper wires act as contact arms for the switch. They rest in the grooves in the bolt head to close the circuit. This double pole double throw (D.P.D.T.) switch can be used for instant switching from one piece of electrical apparatus to another in low voltage electrical experiments.

# Inexpensive B Eliminator Runs Auto Radio

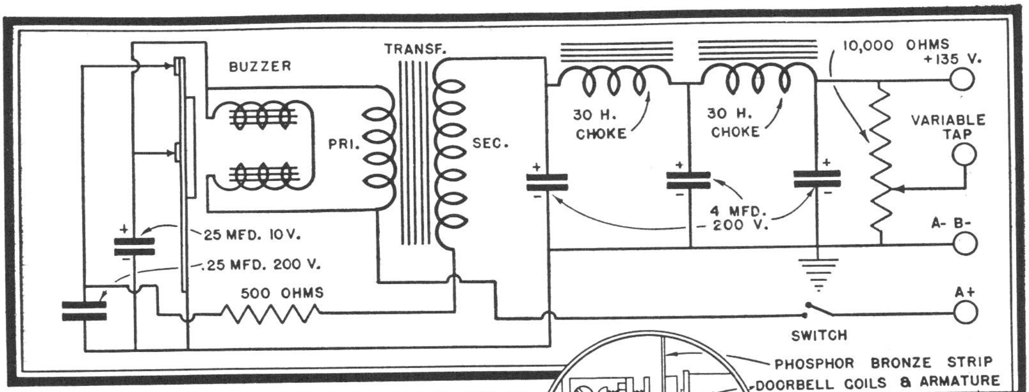


Circuit diagram for complete auto radio B eliminator unit. Old audio transformers with primary and secondary connected in series will do for filter chokes. Filter condensers may be electrolytic, in one unit of 4-4-4 mfd., 200 v. rating. Ground metal case of vibrator to reduce radio frequency interference set up by sparks from motor.

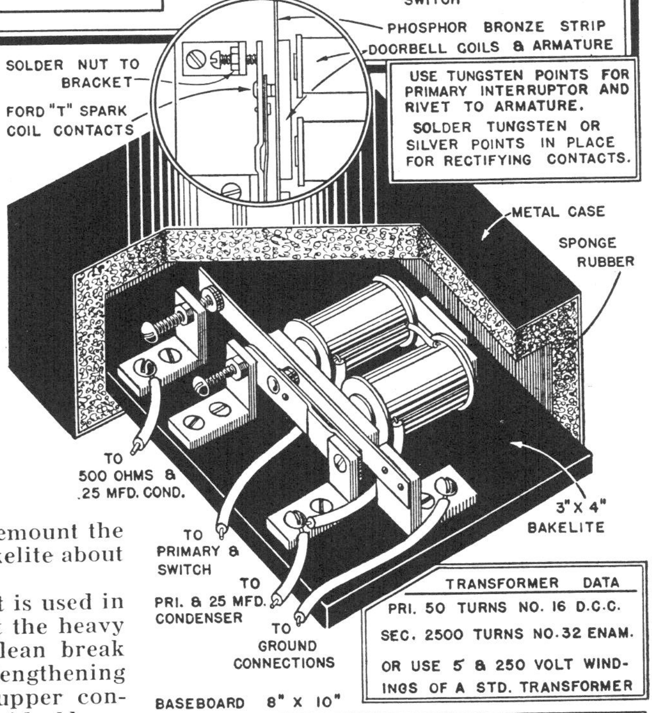


THE construction of the vibrating type of auto radio B eliminator with self rectifying contacts is fairly simple, and requires only parts which are easily obtainable by any radio experimenter. Even if all parts have to be bought the total cost should not be much over two dollars.

The vibrator unit is, of course, the essential part of the eliminator. Take an ordinary door bell or buzzer apart, and remount the magnet coils upon a piece of bakelite about 3"x4" in size.

A hammer break fixed contact is used in the primary circuit to interrupt the heavy 6 volt current. This gives a clean break with a minimum of sparking, lengthening the life of the contacts. The upper contact of a Ford T spark coil will be ideal here.

In order to reduce the mechanical noise of the vibrating contacts, the bakelite panel is mounted in a sheet metal case lined on the inside with 1 inch sheets of sponge rubber. Flexible wires are brought out through holes in the bottom of the box, since stiff leads would transmit some of the vibration.

The transformer is needed to step up the interrupted current. It may be wound, using the core of an old radio power transformer, or may be purchased at a radio supply house. Get a fair size power transformer with a 5 volt winding capable of carrying 2.5 amperes and a 250 volt winding.

If winding the transformer select a core having high grade silicon steel laminations.

BASEBOARD 8" X 10"

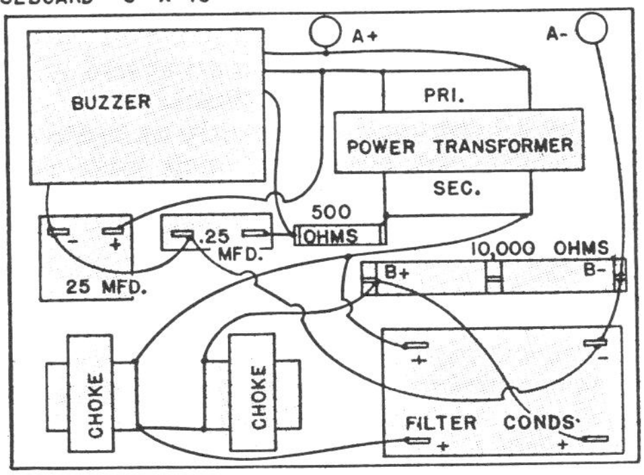


Suggested layout for vibrator, condensers, and filter unit. Mount in neat metal case under or inside auto radio. Use separate leads to battery for B eliminator and for tubes. Eliminator will draw about 2.5 amp. from storage battery.

# Spark Coil 'B' Power Supply Operates From 6-V. Battery

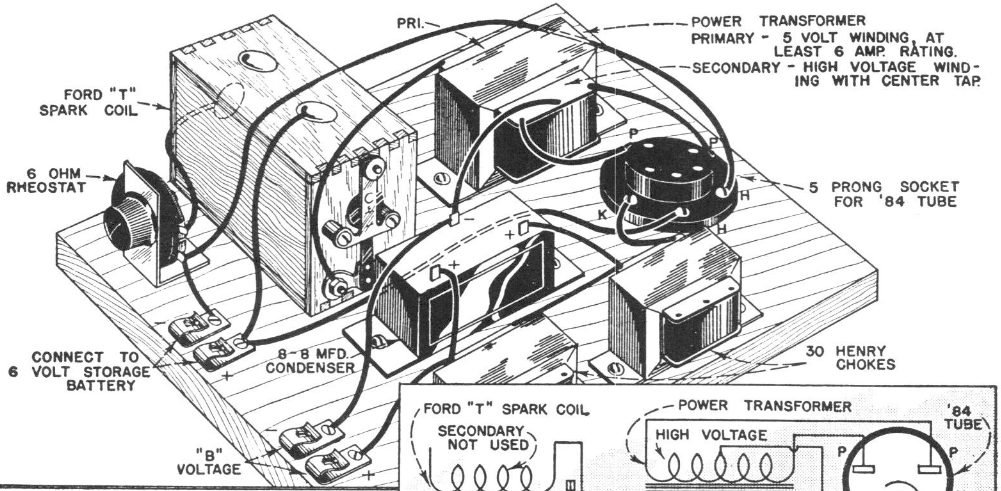


Suggested arrangement of parts for power supply unit. Adjust rheostat and contact points until unit is delivering maximum output. Old audio transformers may be used for filter chokes by placing coils in series.

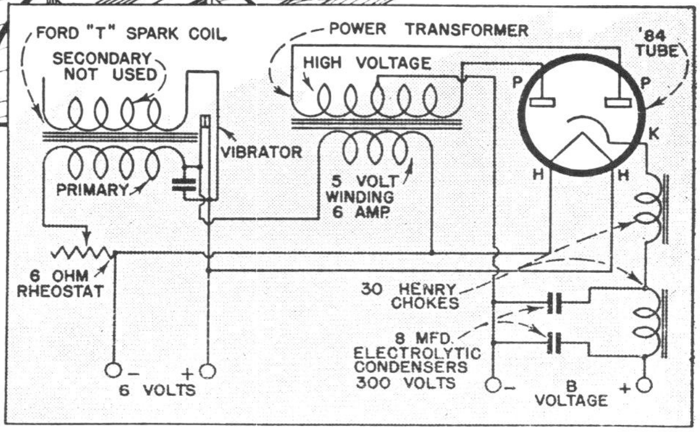


Circuit diagram for vibrator type power supply delivering up to 200 volts.

FOR short wave sets, portable camp receivers, small transmitters, and other low power radio apparatus requiring less than 200 volts this spark coil power supply will be found useful. Operating directly from a 6-volt auto storage battery, it will deliver a practically pure direct current when properly adjusted and filtered.

Get a standard Ford T spark coil; replacement coils often are wound differently and would not work in this circuit. Other parts generally can be found in the radio scrap box; the power transformer should have a 5-volt winding capable of carrying at least 6 amperes, and a center-tapped high voltage winding wound for 600 volts each side of the tap. The turns ratio for the transformer must be at least 7 turns per volt; otherwise the current will be too great and vibrator points will be ruined quickly.

Make all connections exactly as indicated in the sketches, soldering each joint carefully. Mount the entire unit in a wood box, making it as sound proof as possible.

Adjust the vibrator contacts until they give off a high-pitched note, and test with a voltmeter to determine if this is giving the highest output. Keep the vibrator points clean, removing pits and irregularities occasionally with a fine emery stone.

Separate wires should always be run from the storage battery to the filaments of tubes in the radio set, to prevent any pickup which may make the set noisy.

## Safety Pin Makes Clip Connector

A SIMPLE, yet surprisingly efficient, connecting clip can be made from an ordinary large size safety pin as illustrated in the drawing below. The catch is cut off and the ends wound spiral shaped. If you want an insulated handle, flatten the end and force on a bottle stopper. To mount the clip solder it to the head of a brass screw, and attach wires in place by pressing down the cork and inserting wires in the spiral.

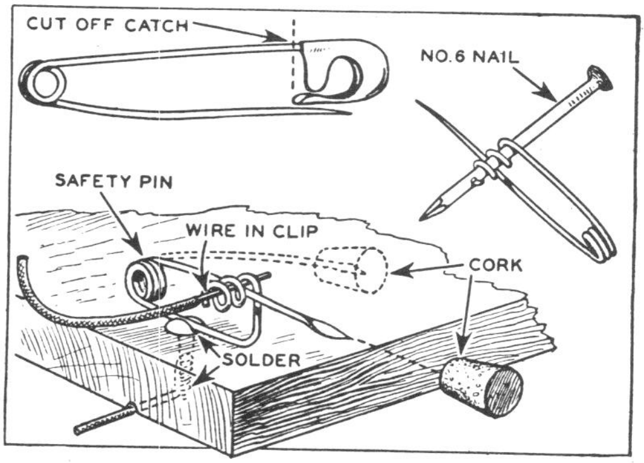


Wire makes positive contact when inserted in the spiral.

# Noise-Reducing Kinks for the AUTO Radio

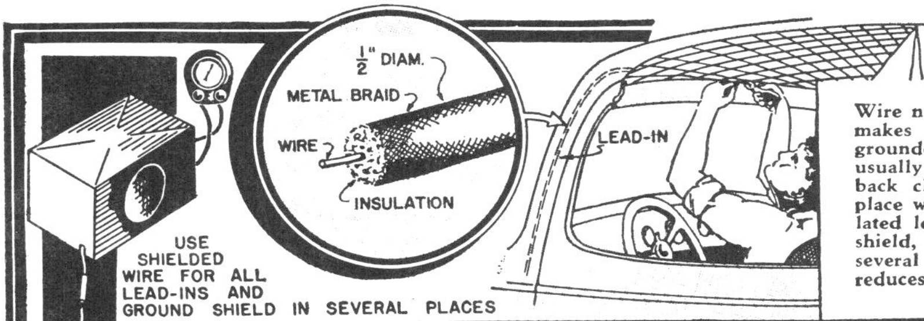


Wire netting in roof of many cars makes excellent aerial if not grounded to frame. Grounds are usually at front and back; peel back cloth, cut wire, and tie in place with strong cord. Run insulated lead-in wire up along windshield, and solder to wire mesh at several points. Shielded lead-in reduces noise, but volume as well.

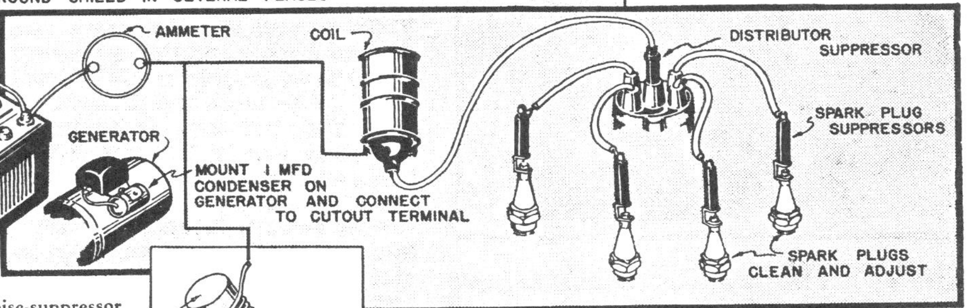


Standard noise-suppressor set includes condenser on generator and suppressors in high tension leads. Connect choke coil in series with dome light wire if roof antenna is used. Keep antenna lead-in as short as possible, and away from all ignition and other wires.

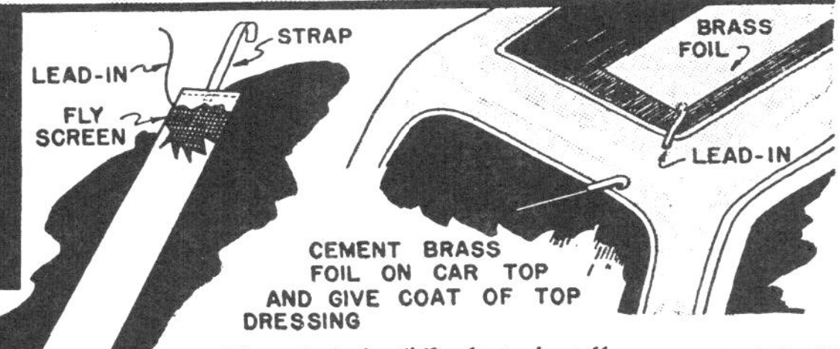


Axle-to-axle screen antenna is made from copper fly screening, 1 ft. wide and 1 ft. less than wheel base length. Enclose in canvas envelope to keep out dirt which might insulate strands. Solder antenna lead-in to screen at closest point to set. Distant stations may be picked up in daytime if lead-in wire is not shielded.

Though it is difficult to install copper screen antenna neatly in roof of closed car, thin metal sheet cemented to top of car and painted with top dressing will be unnoticeable but give good results.

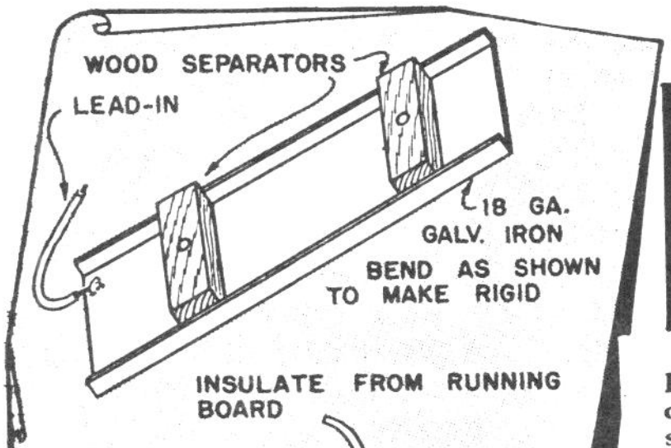


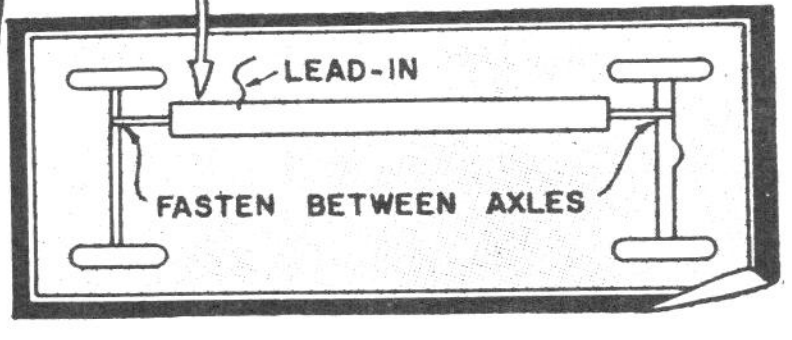


Running board antenna is length of cadmium or copper-plated sheet metal mounted on wood spacers underneath each running board. Connect together with heavy insulated wire.

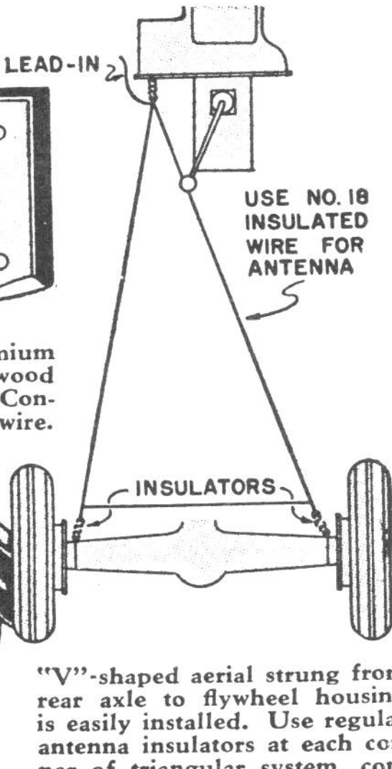


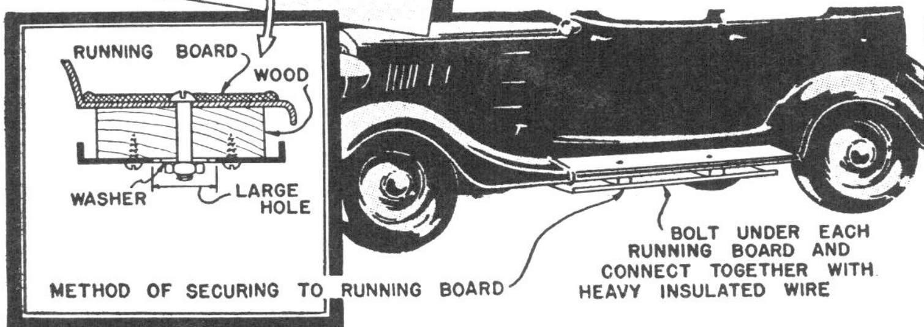


"V"-shaped aerial strung from rear axle to flywheel housing is easily installed. Use regular antenna insulators at each corner of triangular system, connecting lead-in at forward end.

Suppression of ignition noises while engine is operating is biggest problem of auto radio installation. Standard suppressor kits are satisfactory for most cars, and are easily installed. Where noise still persists, above kinks should clear up trouble.

# Powerful 5-Meter Radiophone

**Here's what M-M radio enthusiasts have been clamoring for—an easily operated voice transmitter for talking around town to radiophone stations of fellow fans. This remarkable little outfit works down on 5 meters and can be built from old discarded broadcast receiver parts.**

Fig. 1. Here is most efficient arrangement of parts on baseboard. Tandem condenser is mounted on bakelite panel.

By
THORNTON HALLETT

FOR several months now radio amateurs have been using the much neglected five meter band (56,000 to 60,000 kilocycles) for local phone work and directional transmission. This particularly because the 80 and 20 meter bands are closed to all except those who pass the special examination for operation in those two bands.

Five meters is exceptionally suited for local work, its effective range being in the vicinity of 15 miles. Beyond that its efficiency drops off rapidly.

The transmitting apparatus is simple in nature, the antenna system is small and its width in kilocycles is wide, 4000 kilocycles, thus allowing many stations to work without interference. The power can be very low and receiving tubes used for all work, both key and phone.

### 5 Meter Band Ideal for Beginners

Being open to beginners with radiophone this band offers an ideal field to experiment because no interference will be felt from stations just a few miles off and neither will you interfere with them.

The following transmitter will operate nicely within the limits of this band. It is not claimed to be perfection because apparatus for these high frequencies is far from perfected, but it *does* oscillate nicely and in many tests has transmitted well modulated phone signals over ten miles or more with the tubes and power shown.

The transmitter can be entirely constructed of receiver parts and gives the amateur a new target to shoot for in the way of transmitting equipment. Breadboard style of construction has been followed throughout because of ease of adjustment, and no particular "low-loss" methods used. But it works and that's what we are after. So go to it boys. Let's see what you can do with 5 meters.

### Circuit Simple But Efficient

The circuit employed is the conventional TNT or push-pull, now so popular with amateurs. Use any suitable base about 10x18 inches. At the rear assemble a binding post strip to include B plus, B minus, C plus, two A posts, C minus, two posts for key and two for the "mike".

# Uses BROADCAST SET Parts

Follow the layout of the oscillator section about as shown in Fig. 2 and 3, because it should be as symmetrical as possible to aid in balancing the operating characteristics. That is, plate leads should be about the same length, etc., so as not to introduce an unbalanced condition.

As Fig. 4 shows, two sockets are placed with filaments in parallel. Connect the plates to the two stator sections of the tuning condenser. The rotors of this same condenser are joined together and connect to B minus.

Use a gridleak of about 10,000 ohms and connect it in the center of the wire bridging across the two R.F. chokes in the grid circuit. Bridging these chokes is the grid coil. This consists of 4 turns of No. 14 copper wire wound an inch in diameter with the turns about ½ inch apart. The R.F. chokes consist of 18 turns of No. 30 DCC wire wound on a film spool, turns 1/16 inch apart. With a match dowel, glue the spool to the base.

Connect the stators of the tuning condenser to the ends of a 4/5 turn of heavy copper wire or tubing 3 in. in diameter supported in a rigid position. Then at its exact center solder a wire leading to a third R.F. choke, a milliammeter reading from 0 to 100 and to a dropping resistor connecting to

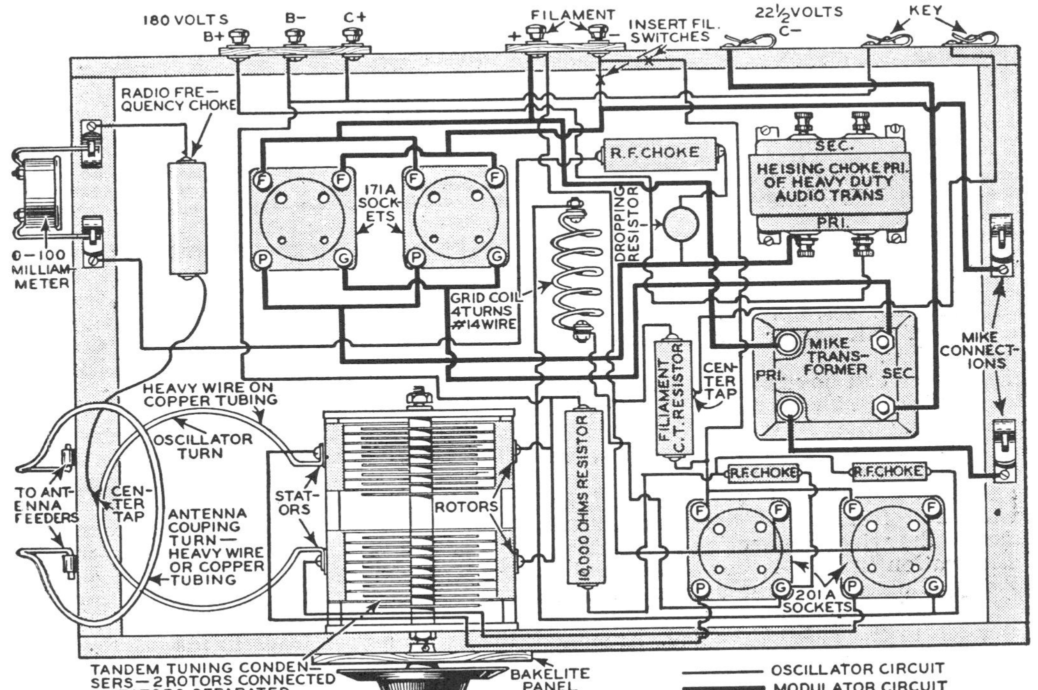


Fig. 2. Here is how various parts are wired up on baseboard. Plate, grid, and filament leads of circuits should be of same length for proper balance of circuit. R. F. choke between dropping resistor and milliammeter should be eliminated.

## Parts for Transmitter Can Be Taken From an Old Broadcast Receiver

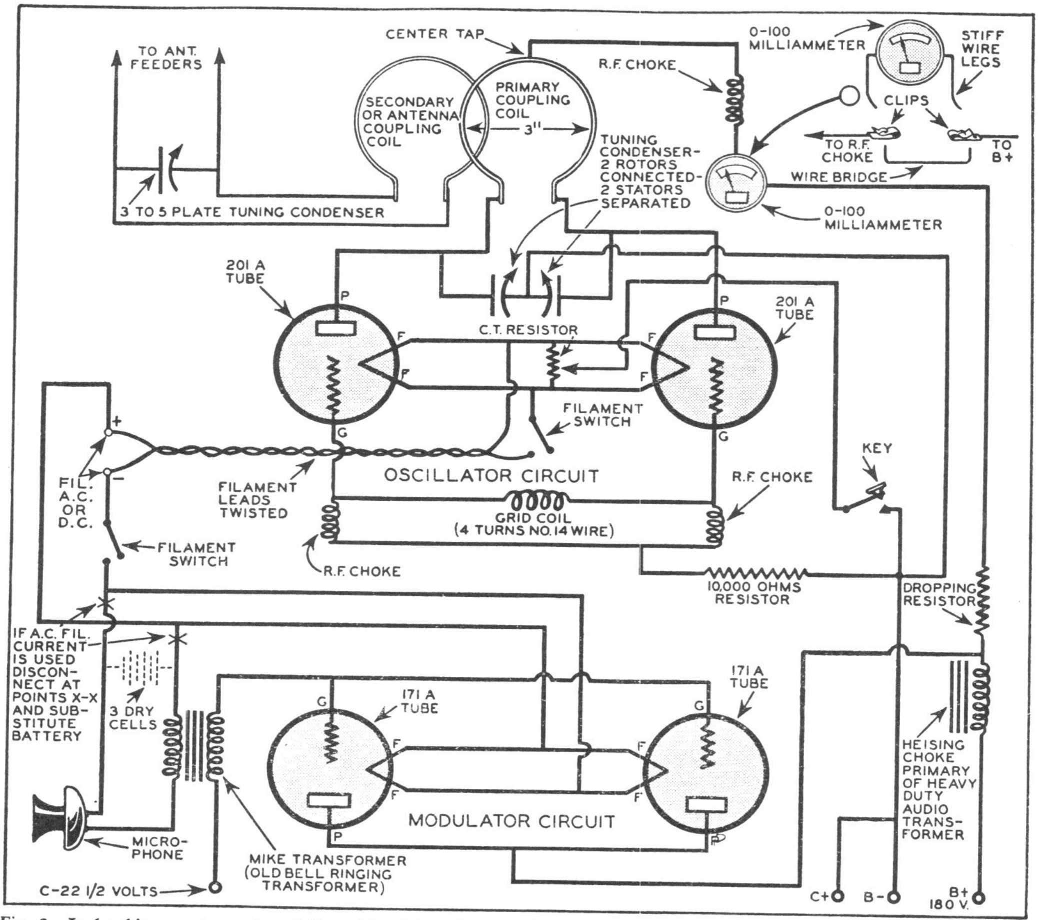


Fig. 3. In hooking up transmitter follow this wiring diagram along with Fig. 2. Note how milliammeter may be inserted or removed from the circuit. Output (8 volts) circuit of mike transformer hooks to microphone; input (110 volts) goes to C minus and 171A grids. If D. C. filament current is used hook up mike as shown in this wiring diagram.

the output of a 15 henry, 100 mil audio choke.

This latter can be the primary of a heavy duty audio transformer, if small tubes are used. A "B" eliminator choke will do for heavier tubes.

The meter can be made removable from the circuit by fastening stiff wire legs to it and inserting these in two clips in the "B" plus line, bridging them with a wire when the meter is removed.

The filament circuit wires should be twisted and fastened to the symmetrical center of the wires connecting the sockets, as illustration in Fig. 4 shows. Connect a 100 ohm filament resistor across the filament circuit at its center and lead the center tap to one side of the key. Arrange a switch for operating the filament circuit as shown. The other side of key connects to the gridleak (10,000 ohms resistor) and B minus.

The modulator section can be arranged at the right end of the baseboard in any convenient way (See Fig. 1). Wire two sockets in parallel, with filaments, grids and plates connected up in pairs. Connect the plates to the output of the Heising 15 henry choke as shown.

Then connect the input of this choke to B plus 180 volts. C plus post connects to B minus. Filaments of these tubes connect in parallel to the filament source, also provided with a switch.

The microphone transformer can be a regulation single button affair or can be home made, as illustrated in Fig. 4. One good instrument can be made by using an 8 volt bell ringing transformer. Connect the 8 volt output leads to the microphone, and the grids and C— to the 110 volt input side.

Also an old audio transformer can be used. Remove the core, dip coil in hot water and push out the primary. Wind a second primary of 250 to 300 turns of No. 28 or No. 30 wire. This will make a good mike transformer, connecting the primary to the mike circuit.

Microphone voltage can be taken direct from the modulator filament circuit if DC current is used for tube lighting. If AC is used disconnect the mike circuit at points X-x and make it a separate circuit, using 3 dry cells or a storage battery at dotted lines (See Fig. 3). Disconnect the mike from the battery when not in use.

The antenna coupling coil is shown in detail. This consists of a duplicate of the oscillator coil but with legs bent down at right angles and out laterally to rotate in a pair of Fahnestock clips fastened to the side of the base. This is shown in sketch and photos. The antenna feeders are attached to this.

Using a 180 volt B eliminator for power (connected to B— and B+180v) and a 5 volt battery or transformer for filament current, the dropping resistor should be sufficient to drop the B voltage to the oscillator tubes to about 135 volts.

About a 2 watt, 800 ohm resistor should do this using 201 A tubes as oscillators. Using two 171A tubes as modulators at 180 volts use about 40 volts of C bias to gain good quality modulation.

**Choice of Three Antenna Systems**

Three antenna systems are shown in Fig. 4 top. All are efficient so you can use whichever one suits your location best. The system illustrated at left of Fig. 4 uses a vertical antenna and ground. The vertical portion is about 12½ ft. high fastened overhead

*(Continued on page 50)*

Front view of transmitter. Oscillator tubes (201A's) are at right front, modulator tubes (171A's) are at left rear. Power, filament, key and mike, connections are on rear.

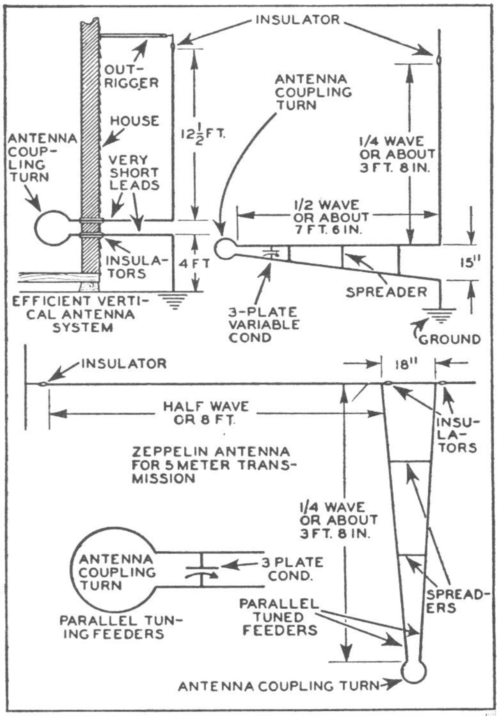


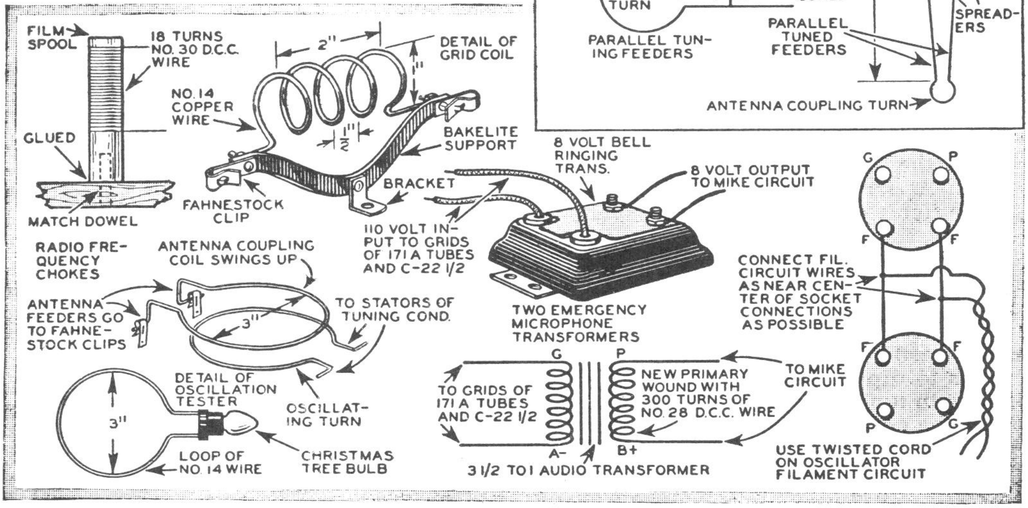


Fig. 4. Top right drawings show most efficient 5-meter antennae for use with transmitter. Lower drawing shows construction of various parts. Grid coil is supported by bakelite "V". Note symmetrical wiring on filament circuit at right.

# POROUS CUP Ground Eliminates CRACKLING NOISES

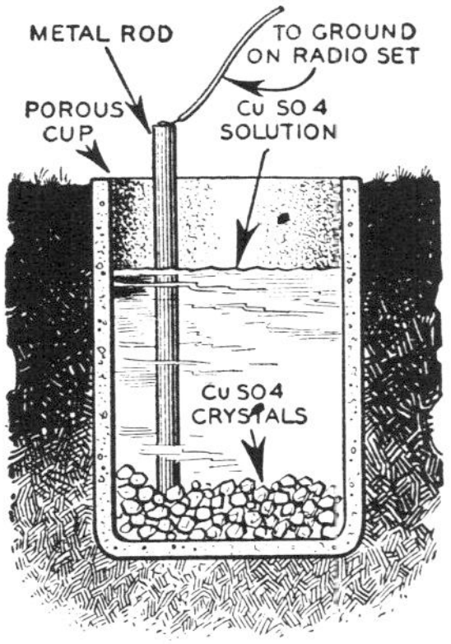


GALVANIC action between iron rod and ground sometimes causes ordinary grounded radio sets to emit terrible crackling noises. The difference in potential between rod and ground may sometimes be as high as 1,000 millivolts, and varies with electrical or magnetic disturbances in the earth.

A porous cup electrode buried in the earth, with no metallic part of the ground circuit touching the earth at any point, will entirely eliminate these crackling noises. Fill the cup with a concentrated solution of copper sulphate, and make contact to an iron or copper rod inside the cup. The copper sulphate seeps slowly through the cup into the ground, making excellent contact. The cup should be filled occasionally with the same solution. The resistance of this type of ground is usually around 400 ohms, but varies with the kind and moisture content of the soil. A ground of this type sets up absolutely no galvanic action, eliminating the problem of varying potential between rod and ground, and greatly improving radio reception.

## Powerful 5-Meter Radiophone Uses Broadcast Set Parts

*(Continued from page 49)*

to an outrigger and insulator. Ground wire should be about 4 ft. long to point of grounding. Make leads as short as possible to the coupling turn on the transmitter.

System at the right is similar but the feeders can be ½ wave long or 7½ feet and the vertical portion only 3 ft. 8 in. high.

In the lower system we have the regulation Zeppelin type with a half wave or 8 foot flat top. One feeder is dead ended at the insulator. The other feeds the antenna. Make these ¼ wave long or about 3 ft. 8 in.

With either arrangement using feeders, spread the latter out to a distance of at least 18 in. at the far end to take care of leakage losses occurring in feeders at these high frequencies. If the feeders need tuning use a 3 to 5 plate variable condenser across the feeders near the coupling turn as indicated in Figs. 3 and 4.

With the antenna feeders attached to the coupling turn, raise the latter at right angles. Insert the milliammeter in the B plus oscillator circuit and turn on the filaments of that portion.

Press the key and watch milliammeter needle. At its dip in the scale bring the coupling coil down about 2 inches above the oscillating turn. This should raise the meter reading somewhat if the antenna is cut near the frequency the oscillations occur.

If the feeder tuning condenser is operated this upward reading should increase until it starts to dip again. At this peak the antenna is in tune with the oscillator. Do not couple the coils too close. Experience will show this best distance for good coupling.

To determine the limits of the five meter band may prove difficult. The best way, of course, is to find someone with a five meter receiver to work with you. He tunes his receiver to the band limits and monitors your transmitter from a distance of a few hundred yards. When this is done mark your dial readings.

When the oscillator is properly tuned and the antenna in resonance with it, then turn on the filaments of the modulator tubes and the microphone circuit. Talking in the mike will modulate the output of the oscillators and the voice will be heard distinctly, even over a twenty meter receiver heterodyned with the carrier of the five meter job.

To prove oscillation, use a single turn of wire about three inches in diameter connected to a flashlight bulb (See Fig. 4). The coil and bulb should be brought down to within a couple of inches of the oscillating turn. If oscillation is taking place the bulb will glow.

The photographs show the rotors of the variable tuning condenser containing more plates than can be used for five meter work. After the transmitter was used and oscillation proved, one plate at a time was removed from each rotor until finally three plates were left in each. This caused the set to oscillate in the five meter band.

The peculiarities of five meter transmission are many but will prove interesting. Hills, high buildings, woods, etc., will often prove barriers to such signals at certain periods of the day and prove ineffective at others.

By tuning each of two transmitters to opposite ends of the five meter band two way conversation or "duplex" can be carried on. This means that each transmitter can be kept on the air and thus questions and answers passed exactly as on a land telephone.

# Build 'Master Mike'—An Efficient Condenser Microphone

THIS efficient little condenser microphone, with a cost varying from nothing to fifty cents, will reproduce voice or music through the parlor radio, and at the same time give to the amateur radio enthusiast a mike compatible with those used by the broadcasting stations. It has a far wider responsive range than that of the cheap home mikes with which the market is now flooded.

An old radio headphone forms the case of this mike. Remove diaphragm and magnet coils, and drill a ¼" hole in the center of the case. Cut off an ordinary wood thread spool so it will project slightly above the rim when set inside the case. Carefully center this spool and fasten in position with small bolts running through the spool flange and the back of the case.

### Drilling the Rigid Plate

To make the rigid plate of the mike, draw a circle with diameter ¼" less than the inside diameter of the case on an old condenser plate, and carefully cut out with coping saw and file. Drill a small hole in the exact center of the disc, and drill other holes of about 1/16" diameter over the entire plate at regular intervals.

Glue a piece of coarse sandpaper to a flat piece of wood somewhat smaller than the case, and carefully dress down the top of the spool until the aluminum plate, when placed in position on the spool, will be no more than 1/32" below the level of the case rim at any point. Check this carefully by holding the case up to a light and passing a try square over the rim.

Select a small diameter bolt about ¼" longer than the depth of the case. With a drill slightly smaller than the diameter of this bolt make holes in two small corks. Screw these corks on the bolt, trimming the last cork put on so it will just fit in the ¼" hole in the case, and making the first one big enough so it will not pass through this hole. These corks act as insulating bushings and at the same time provide a means for adjusting the fixed condenser plate. Push a third cork bushing into the spool from the top, to act as a centering washer.

### Aligning the Rigid Plate

With the terminal bolt in place, ream out the center hole of the aluminum diaphragm until the terminal bolt can just be forced into it. This joint should be tight, and may be soldered if superfluous solder is removed with sandpaper.

Tighten the large cork on the back of the mike case, to draw the diaphragm down to the spool.

Enlarge the hole in the bakelite earpiece to about the size of a half dollar, using a

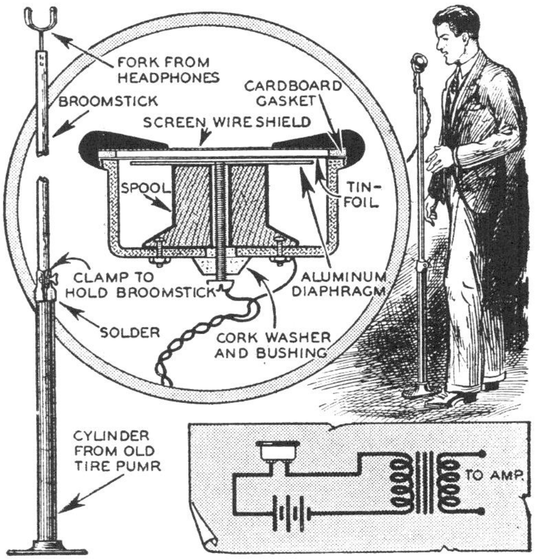


coping saw. First with a round file, and then with sandpaper, smooth and round out the rough edge.

Next, cut a round piece of ordinary screenwire that will just fit inside the earpiece. This forms a diaphragm shield to protect the easily damaged tinfoil. From cardboard, cut a narrow ring-shaped gasket that will fit between the screen and the vibrating diaphragm at the edge.

### Stretching the Tinfoil Diaphragm

Great care should be taken in stretching the vibrating diaphragm, which is made of ordinary tinfoil from chocolate bars, or any metallic foil which tinkles when handled. See that it is absolutely free from wrinkles before the stretching process is begun. Then stretch the foil firmly over the rim of the case, pull it free from wrinkles and folds in all directions and slip a rubber band around the top of the case to hold it. With the diaphragm shield and cardboard gasket in place, slip the earpiece over the case and screw down slowly. To prevent tearing the diaphragm, it might be well to give the foil a coating of petroleum jelly where the threads of the earpiece come in contact with it. Continue to tighten the earpiece until the foil is taut.

Hook the unit in series with a small battery across the input binding posts of any audio amplifier of two or three stages. Quite a bit of experimentation may be necessary to get the proper battery voltage for your mike, since the voltage needed will vary with the size and capacity of the unit. When the amplifier is turned on, speak into the mike. If the sound fails to reproduce, it may be because the diaphragms are too far apart.

# Ultra-Short Wave Receiver

By THORNTON HALLETT

**Here are the plans for a good three-tube receiver which picks up the 5-meter voice signals. With these two inexpensive outfits rigged up together you will have an ideal radiophone station that will introduce you to the keen thrills to be had from talking over the air.**

THIS is a companion to the five-meter transmitter previously described. Like the former, it was put together experimentally and therefore the layout need not be explicitly followed. It was constructed entirely of spare receiving parts and, exclusive of the tubes, should cost the builder very little.

Let it be understood that much work is still to be done on five meters and other types of receiver circuits can be used, but the super-regenerative appears to be the best for those frequencies.

### Receiver Easy to Operate

Good tubes must be used. Some may not respond to these high frequencies but, by trying several, you will have no trouble to get the outfit working. It acts similar to the ordinary regenerative receiver with one exception. When looking for signals the noise level will be high. But when a signal is crossed the noise will die out and the signal come in clear and sharp.

A good vernier dial *must* be used on the tuning dial and the tuning condenser also must be one with a quiet rotor connection. The resistors also should be extremely quiet in operation and for this reason IRC resistors were chosen for this particular job.

### Arranging Parts on Baseboard

The baseboard used measured 7 inches across the front and was 12 inches deep. By looking at the baseboard layout diagram (Fig. 2) you will see how the various parts were assembled. Slight alterations of this assembly will have no serious effect on the receiver's operation.

First place the .0001 variable tuning condenser in the center of the front with space enough for the R. F. choke in front of it. A piece of rubber or cigarbox wood fastened upright to the baseboard edge will suffice to support the vernier dial.

At the left of the tuning condenser is placed a phone jack or phone tip jacks while at the right is fastened the antenna tuning condenser. This latter consists of two ½-inch square aluminum plates, with right angle base lugs, fastened parallel

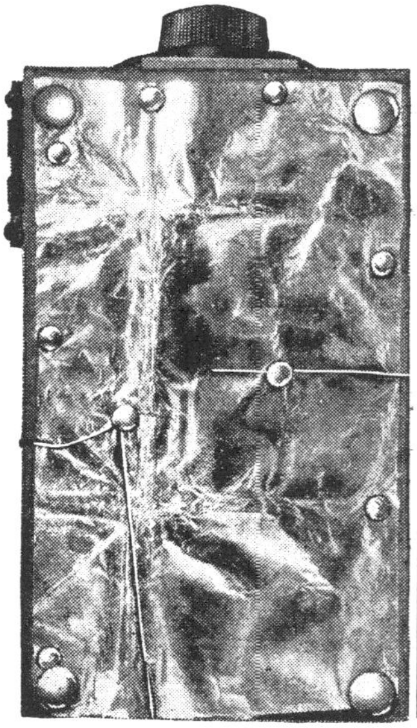

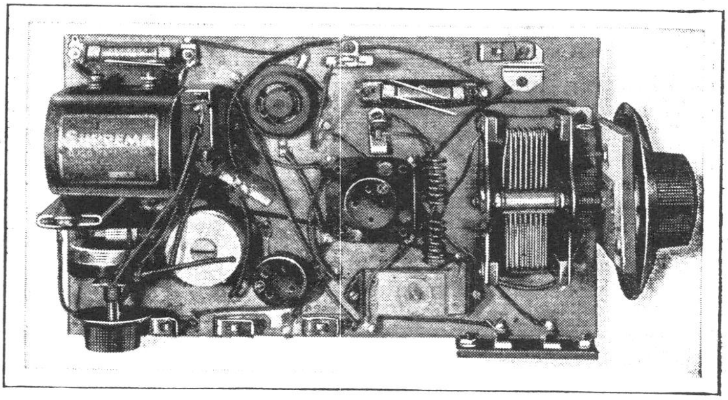

Fig. 1. Bottom of receiver is shielded with coating of tinfoil as shown at left. All grounds in circuit run with direct short leads to this foil. In top view of set, at right, is seen arrangement of parts. Study this in conjunction with drawing Fig. 2.

# for Radiophone Transmitter

**Blueprints for Modern Mechanix 5-meter Radiophone Station**

**A set of large-size shop blueprints for complete radiophone station, including the 5-meter transmitter shown on page 46, and the supplementary 5-meter receiver shown herewith, all made from original drawings, are available for $1.75 from the Blueprint Department of Modern Mechanix and Inventions, 529 So. 7th St., Minneapolis, Minn.**

**How to Become a Radio Amateur**

**This little 5-meter radiophone will initiate you into the radio amateur's world. However, you will encounter some difficulties, like mastering the code, learning radio fundamentals, and obtaining your government license. How to overcome these difficulties with greatest ease is told in a little booklet, "How to Become a Radio Amateur," available here at 25c.**

about 1/16 inch apart. A Fahnestock clip is connected to the outside one.

The three sockets are placed as shown in photo and diagram, the middle and left one being 4-prong affairs while the right socket is a five-prong or UY.

Just behind the tuning condenser and at the left is placed a midget condenser of about .00001 capacity. This can be a rotor type if handy but the writer used a book type balancing condenser with equally good results. It was taken from an old broadcast receiver. At the right of the base is placed a holder and a 2 meg. gridleak.

The auxiliary oscillator and audio system is at the rear end of the baseboard. This consists essentially of any good audio transformer, a 50,000 ohm variable resistor, preferably of the carbon type to avoid noise, the auxiliary oscillator coils, a 150,000 ohm resistor across the secondary of the audio transformer and the five fixed condensers shown.

Fahnestock clips can be placed at convenient points along the edge of the base for connections to A plus, B minus, B plus 90, B plus 45, B minus and Ground.

Now turn the receiver over and shield the bottom of the base board with heavy tinfoil as shown in the photograph (Fig. 1). This was done by shellacking the wood and foil one coat and then a second when the first was dry.

While the second coat is still tacky,

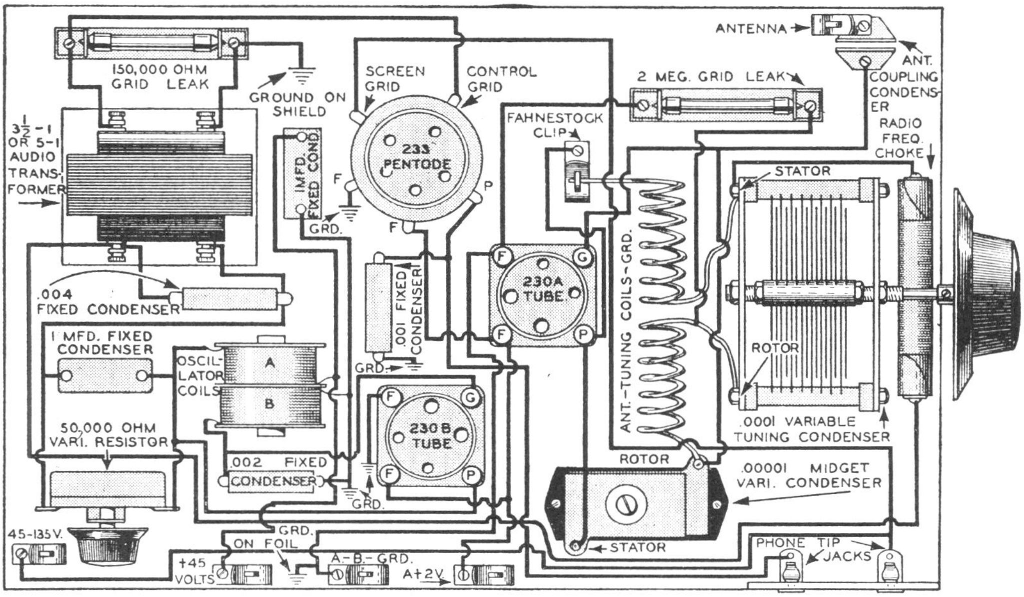


Fig. 2. Arrange parts on 7x12 in. baseboard and wire them up as illustrated. Leads ending in "grd" connect to foil on bottom. Tuning coils are supported by terminals of tuning condenser, which is in turn mounted on bakelite or hard rubber panel.

# *Most Parts for Outfit Can Be Obtained From Old Broadcast Receiver*

Fig. 3. Here is side view of the receiver, giving clear idea of arrangement of condensers, tubes, audio transformer, resistor, phone tip jacks, and Fahnestock clips.

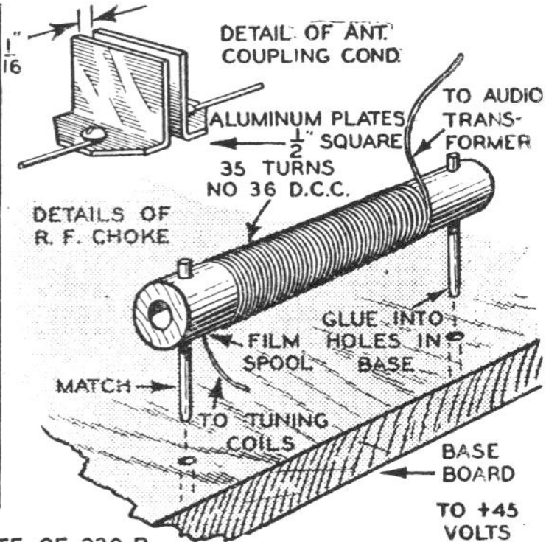


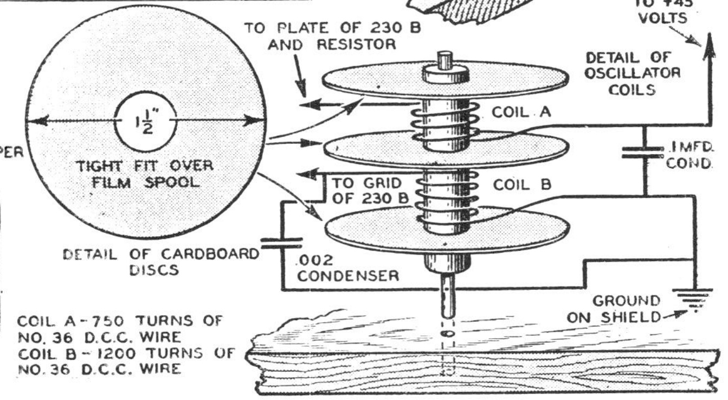


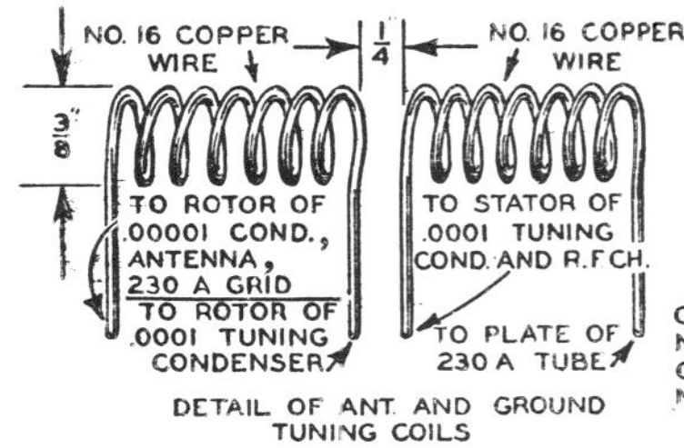


Fig. 4. Parts of receiver which can be made in workshop are shown here. Antenna condenser, seen at top, is simply two bent aluminum plates set close to each other. In details of oscillator coils, at right, coil B should have both terminal connections and winding direction reversed. Top terminal of B goes to .1 mfd. condenser, lower terminal to grid of 230 B tube.

spread the foil smoothly over the wood. A few thumbtacks will also help in preventing the corners from being turned down.

This shielding is represented in the diagram by the ground symbol. Be sure no screws, etc., project through the wood to touch the foil. Four rubber chair legs feet at the under corners of the base serve to keep the shielding clear of the table when the receiver is righted.

A thorough study of the wiring diagram (Fig. 5) will suffice to guide the builder without much additional instruction. However, be careful to make all grid leads as short as possible. In fact all wiring should be made direct and to the point using as little wire as necessary.

### Use Extreme Care in Wiring

Solder all necessary joints carefully and use either insulated wire or bare wire and spaghetti. Run wires from the A minus, B minus and Ground clips directly through the baseboard and connect to the shield. This also applies to all grounds shown in the diagram.

If desired, a switch can be placed in the A minus lead between the clip and the shield for turning on and off the filament current. Wiring to the shield also applies to the negative leads from the tube sockets.

If necessary, the .00001 midget condenser can be dispensed with as it merely serves to center the tuning range on the dial of the tuning condenser. But it is handy and try and use one if possible.

Now as to the construction of the few special parts shown in Fig. 4. The R. F. choke can be easily made by using a wooden film spool or dowel of similar proportions. Drill a small hole through each end into which glue a match stick. Then wind on 35 turns of No. 30 DCC wire, binding the turns with Duco cement. The matches can then be glued into holes in the base directly in front of the tuning condenser and the choke thus held horizontal and rigid.

### Winding the Tuning Coils

The tuning coils are also shown in Fig. 4. Wind them of bare copper wire stiff enough to be self supporting. Number 16 or 14 will do nicely. Wind on seven complete turns for each coil making them ⅜ inch inside diameter and space the turns about the diameter of the wire.

They should be coupled about ¼ inch apart behind the tuning condenser and connected as shown in Figs. 2 and 5. If the two coupled end leads can be fastened direct to the rotor and stator of the tuning condenser it will materially assist in holding the coils rigid. Rigidity is important.

The auxiliary oscillator coils may seem to offer a problem but actually their construction is quite simple. Cut three disks of stiff cardboard 1½ inches in diameter as illustrated in Fig. 4. Cut holes in their centers to make a snug fit over a wood film spool and glue them in place on the spool ½ inch apart.

The spool can be about 1½ inch long. Drill a hole through the center of its length and glue in a small dowel projecting from the bottom about ½ inch. In the top section A, scramble wind 750 turns of No. 36 DCC wire, tagging the beginning and ending leads. Then wind 1200 turns of No. 36 DCC wire in section B, being careful to wind in the same direction and tag those ends as well.

The use of a pentode tube in the audio stage calls for the same screen voltage as applied to the plate and please note that this is taken from the battery side of the phone jack to prevent this unnecessary drain from going through the phones. Type 230 tubes are used for the detector and oscillator while the 233 pentode is used in the audio end. This allows the receiver to be operated from a single storage battery cell or from two dry cells with a suitable resistor to drop the potential to 2 volts.

One word about operating a five-meter receiver. If, when you first turn it on you do not hear any signals it will probably be because there are none within receiving range. The transmitting range at five meters, for high power as well as low, seldom exceeds 30 to 40 miles.

Five meters is essentially limited to local work and from five to fifteen miles seems to be the most useful range. A five to ten foot antenna is usually sufficient, but its height has much to do with reception. So operating the receiver from a room high in the house will probably prove most effective.

Use a good ground and fully charged A battery and fresh B batteries to insure quiet reception. A sure test to see if the thing is actually working at the start is to idle the car nearby. The ignition should so affect the receiver that you will know it at once.

With 90 volts on the pentode voice transmission can be picked up on a good speaker, although 133 volts may be necessary from a distant transmitter. For phone reception 45 volts may be found sufficient.

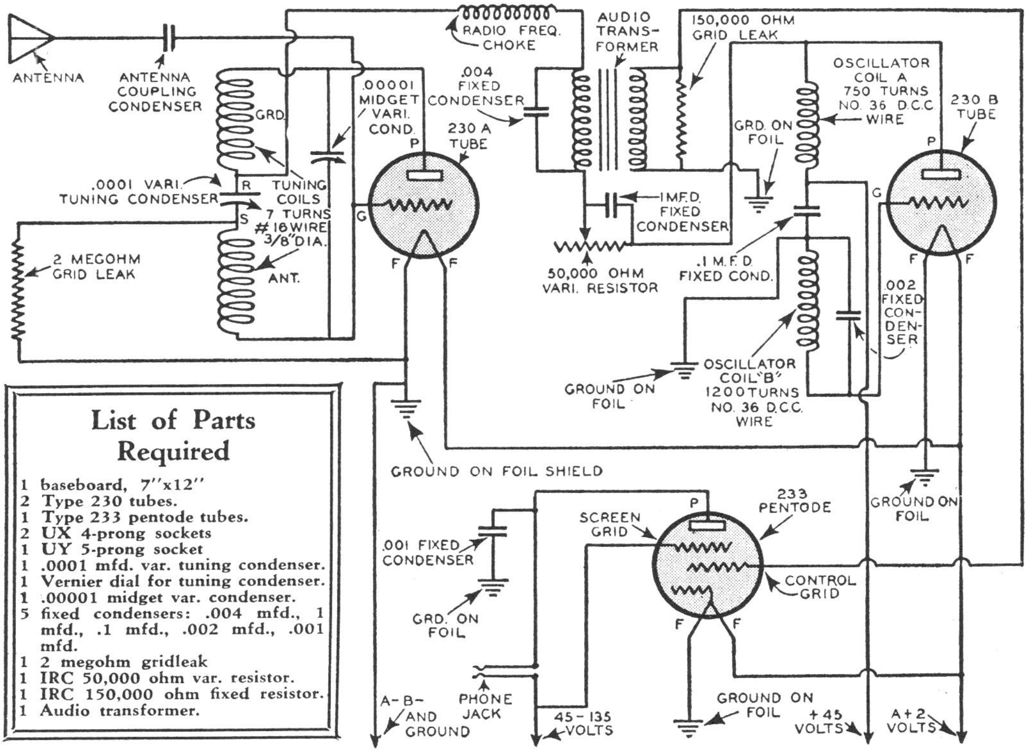


## List of Parts Required

1 baseboard, 7"x12"
2 Type 230 tubes.
1 Type 233 pentode tubes.
2 UX 4-prong sockets
1 UY 5-prong socket
1 .0001 mfd. var. tuning condenser.
1 Vernier dial for tuning condenser.
1 .00001 midget var. condenser.
5 fixed condensers: .004 mfd., 1 mfd., .1 mfd., .002 mfd., .001 mfd.
1 2 megohm gridleak
1 IRC 50,000 ohm var. resistor.
1 IRC 150,000 ohm fixed resistor.
1 Audio transformer.

Fig. 5. Here is wiring diagram of the receiver. Follow this circuit in conjunction with Fig. 2. Phones are inserted in the jacks, seen in lower part of drawing, while tuning is done with the .0001 vari. condenser. Make all leads short as possible.

# ONE-TUBE MONITOR Tunes Your S. W. TRANSMITTTER

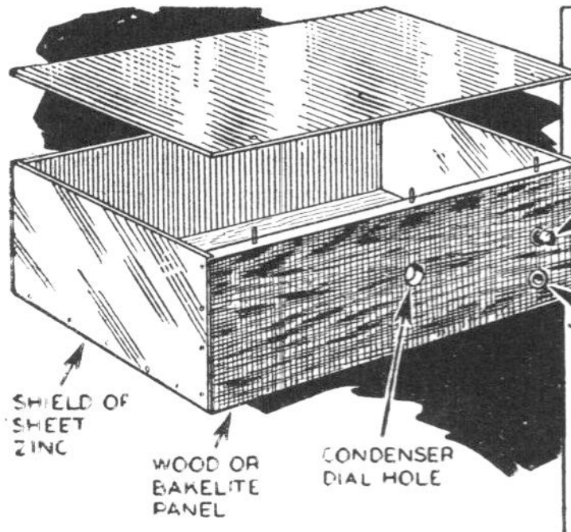


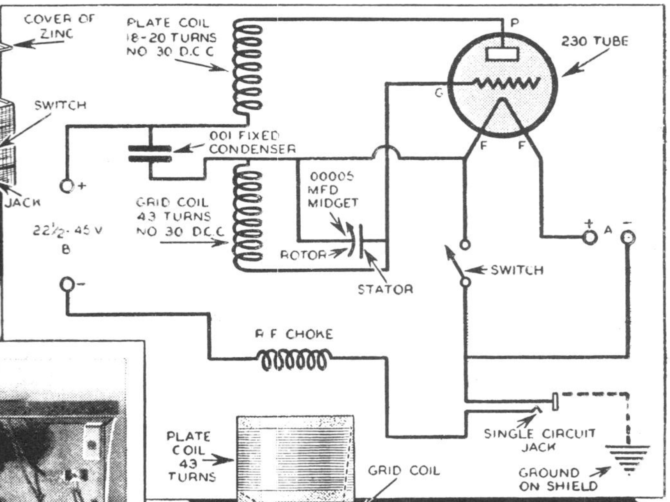


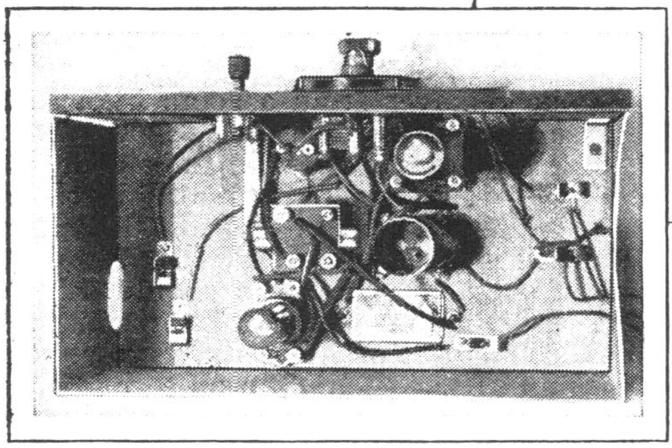

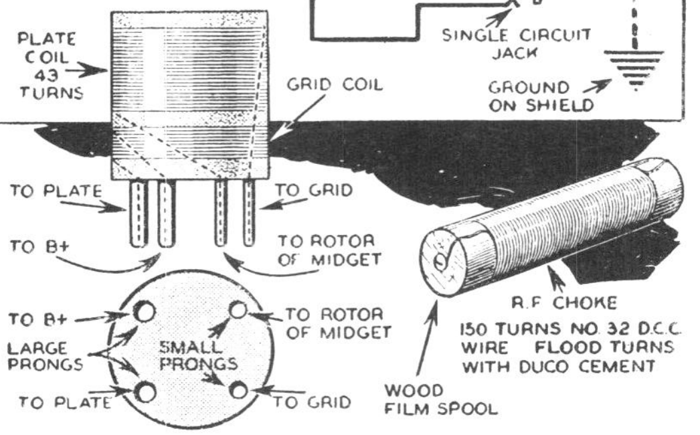


Shield the monitor completely as illustrated above. The circuit to be employed is seen at upper right. Note that one side of phone jack connects to shield or ground. Plate and grid coils are wound on tube base as seen at right.

IN THESE days of narrowed transmitting bands no amateur should attempt to put his transmitter on the air without some means of measuring its frequency within reasonable limits. The penalties are heavy for transmitting outside the boundaries of the prescribed bands so "better be safe than sorry." By the use of this monitor the operator can measure his frequency to within a few hundred cycles. It also serves very nicely as a double check on the wave meter. Construction is surprisingly simple.

The monitor shown in the photo is exactly the same hookup as in the sketch with the addition of a stage of audio. The latter is not necessary, however, and was only added for greater "kick." The base is 12 in. by 6½ in. to allow the insertion of A and B batteries if desired.

This coil is for the 80-meter band, but readings for all other bands can be read by its harmonics which have a good kick even down to 10 meters. To test out the range of the coil, set the monitor to oscillating and run a wire a few feet long from the stator of the .00005 condenser. This acts as an antenna and, with headphones in the jack, should enable you to pick up code stations nicely. A monitor is nothing more than a receiver without an oscillation control.

When completed it should be shielded as illustrated in the drawings. A 45-volt (miniature) B battery is used for plate power and a single cell of storage battery. Two dry cells or even a pair of C batteries can be used for filament battery with suitable resistance to reduce to 2 volts.

To use a monitor to check frequency of a receiver, find a station on the receiver of *known* frequency. Have the receiver oscillating and then with the monitor oscillating, turn the monitor dial until a squeal is heard on the receiver phones.

You may hear several but the loudest one is by all chances the one to use. Turn the monitor dial carefully until you get "dead beat" the point where the squeal reaches a bottom pitch between two higher ones. Mark the monitor dial at this frequency. Do this to other stations of known frequency and after a while you will have several monitor readings of known frequency from which can be plotted a curve.

Knowing these various points on the monitor dial you can check the transmitter frequency. Plug the phones into the monitor and start the transmitter. Set the monitor dial at a desired frequency and then turn the transmitter dial until you hear the loudest squeal in the phones. This shows you the frequency of your transmitter.

# Simple One Tube Oscillator Is Valuable Code Practice Set

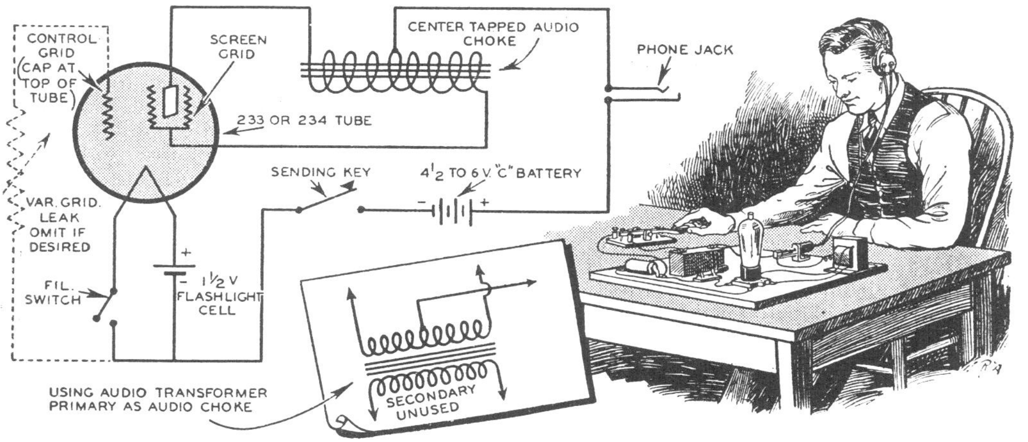


HERE is a simple one tube code practice set that is small enough to be jammed, batteries and all, into the overcoat pocket.

Either a 33 or a 34 tube will serve as oscillator, but the 34 is better, taking far less filament current.

Get a small center-tapped audio choke (center-tapped audio transformer primary will do), an open circuit phone jack, a 5 prong tube socket, a variable grid leak, and a sending key. Connect as shown in the diagram, mounting parts close together to save space. Use a 1½ volt flashlight cell for the filament and a 4½ to 6 volt C battery for the plate supply.

Lowering the grid leak resistance lowers the pitch of the note heard in the headphones, but the set will operate nicely without it.

## Trace Out Transmitter Troubles With Crystal-Meter Test Outfit

THE usual method of tracing out R.F. currents in transmitting circuits or neutralizing buffer stages and power amplifiers is by means of a neon lamp or flashlight bulb in series with a loop of wire held near the inductance. A much more sensitive device can be easily built from odds and ends as shown herein.

It consists of a suitable base or handle upon which are mounted a low reading milliammeter, a crystal detector and and the pickup loop. By holding the loop near the inductance in the usual manner, even the slightest indication is discerned in the meter that would not light the bulb used in the former method.

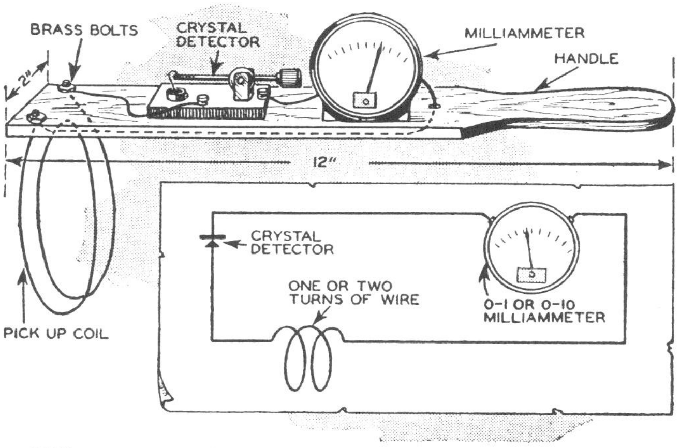


Milliammeter, crystal detector and exploring loop are mounted on the wooden handle as shown. Insert gives wiring diagram.

Fashion a stick to the dimensions given in the drawings. Then make a double turn coil of stiff copper or brass wire about 3 inches in diameter and with the turns a half inch or so apart. Push the end up through holes in the square end of the stick and solder or otherwise fasten to two bolts.

To trace out R.F. currents adjust the crystal detector to a sensitive spot and then hold the coil an inch or two away from the inductance to be tested. As R.F. is picked up the meter needle will rise, depending upon the sensitivity of the meter. When no R. F. flows, the needle stays put.

# Alarm Clock Omnigraph

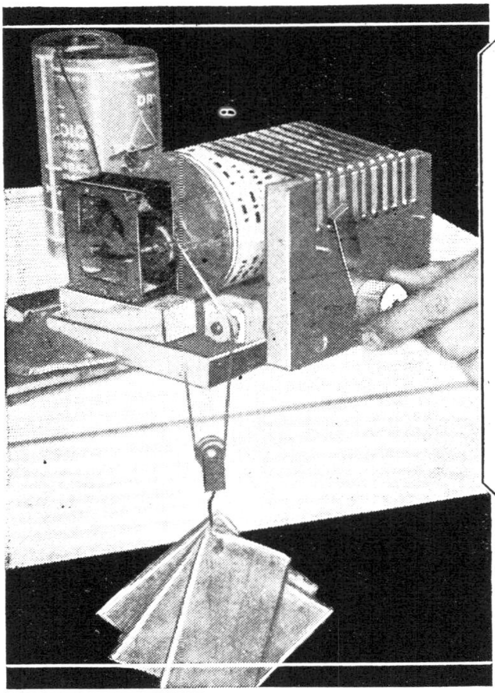

The completed omnigraph, ready for code lessons. Weights suspended on hook turn the drum around which the code sheet is wrapped. The switch arm permits shifting from one set of characters to another in rapid order, preventing anticipation of letters.

In practice lay parts out on table as illustrated above.

**Radio amateurs having difficulty in learning the radio code in preparation to going on the air with a transmitter will find this home-made omnigraph a helpful and patient instructor. It reels off the dots and dashes at any desired speed to develop your proficiency at reception.**

A SIMPLE device to use for learning the radio code can be easily made with a tin can, an old alarm clock works and a few pieces of tin and wood. The omnigraph has been so designed that it is reliable in its operation, but requires no special tools or parts for its construction.

The principal features of the instrument can be seen in the accompanying drawings. The clock works rotate a tin can which is connected to the minute hand shaft. A simple universal joint is used to couple the clock and can.

In preparing the clock works for use on the omnigraph the escapement mechanism must be removed and also the main spring. A small wooden spool is then mounted on the main spring shaft. The clock works is operated by a cord wound around this spool and force applied by means of a weight.

The tin can drum is covered with a paper held in place by rubber bands, as illustrated in Fig. 1. The dots and dashes are cut in this paper and a contact spring makes contact through the holes in the paper as the can is rotated, thus completing an electrical circuit and making the radio signals.

### Operating on the Clock Works

The hair spring and adjacent parts should be removed from the clock works, but the escapement wheel left in place. The wheels will then spin rapidly, depending on the force applied to the main spring shaft. If desired, the alarm spring and wheels and hour hand wheels may be also removed.

The main spring is removed because there is no simple way to regulate the speed if it is used. By substituting a small pulley spool in place of the main spring and driving it with a cord and weight, the speed can be regulated by the weight applied.

This method also gives a longer run per winding than

**How to Become a Radio Amateur**

When you've learned the radio code from this omnigraph you're ready to take your government license and get on the air with your transmitter. How to achieve the status of full-fledged amateur and how to build your own long distance transmitter and receiver are told in the booklet, "How to Become a Radio Amateur," available for 25c from Modern Mechanix Radio Editor, 529 South Seventh Street, Minneapolis, Minnesota.

# Teaches Code to Radio Fans

with the spring. The time is doubled by adding the movable pulley, as shown in the drawing, and doubling the weight. The cord is wound up by turning the winding key in the usual manner. The can should rotate about once in 4 to 8 seconds.

A universal joint between the clock works and the tin can drum is desirable because with the simple mounting used it is difficult to align the shaft and bearing properly. Contact with the tin can drum is made through the end bearing. This is not a perfect way to complete the electrical circuit, but it is entirely satisfactory.

### Rotating Drum From Tin Can

The contact springs make very good contact with the tin can because of the bright tin plated surface. When selecting a can, choose one without dents and with a bright surface. Soak the paper label off. Cut a circular piece of tin and solder into the open end. The tension of the contact springs should be such that they will move downward about ⅛ in. if the drum is removed.

To make a paper with code lessons, first cut an oblong strip of paper the width of the can and about ¼ in. longer than the circumference. (See Fig. 2.) Cross-line or graph paper is best if available. Place the paper on the drum and mark where the contact springs touch it. Cut the holes in the paper with a small chisel or sharp knife.

### How to Mark Dots and Dashes

The holes should be about ⅛ in. wide. Make a dot 1/20 to 1/16 in. long. (Some cross section paper is made with lines 1/20 in. apart.) Make a dash 3 times as long as a dot, the space between signals (dots and dashes) equal to 1 dot, the space between letters 4 dots, the space between words 8 dots.

Since most letters indicate some other letter when reversed, the paper can be

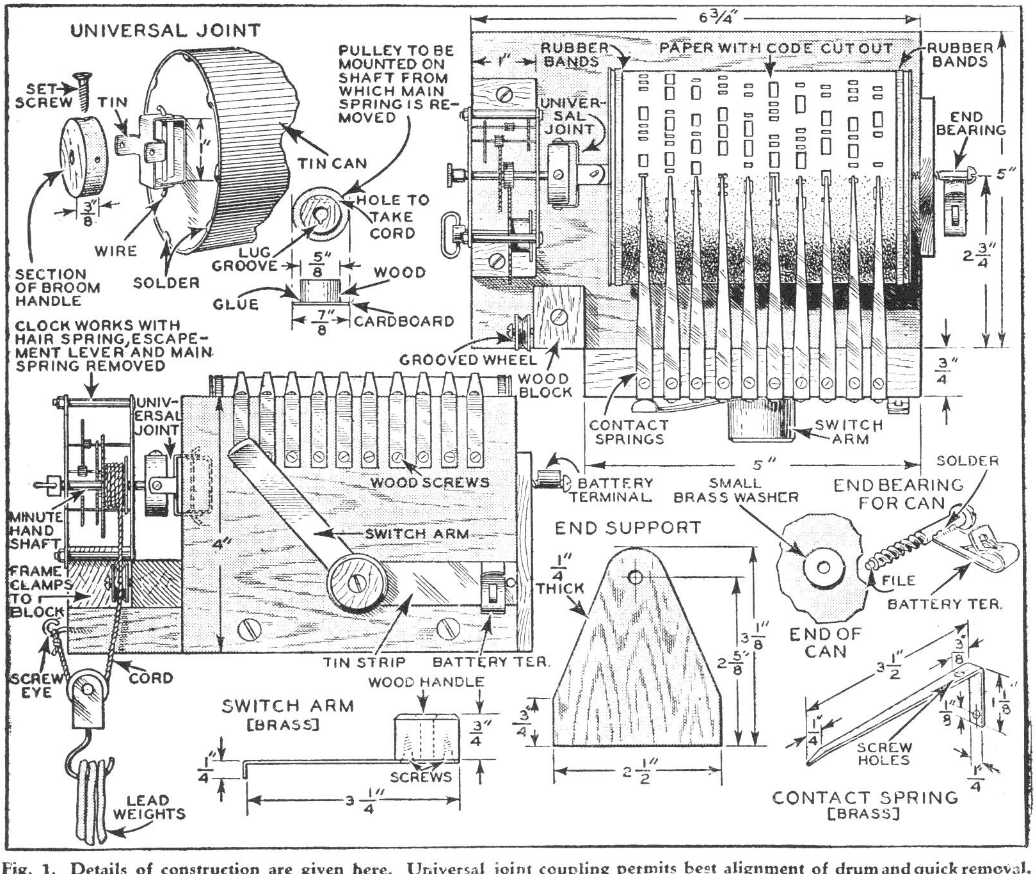


Fig. 1. Details of construction are given here. Universal joint coupling permits best alignment of drum and quick removal. Note how frame of clock clamps to block. Connections to battery are taken off from end bearing and switch arm terminal.

# *Dots and Dashes Cut in Graph Paper to Form Characters for Practice*

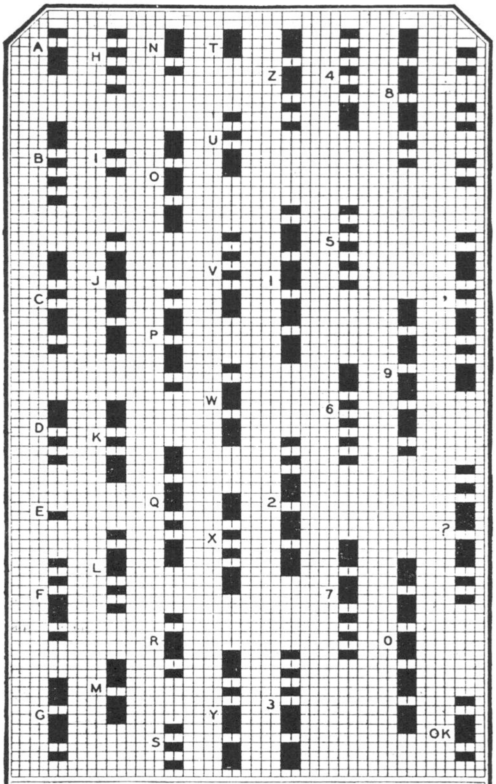


**Fig. 2. Here is how dots and dashes of characters are cut out of squared paper.**

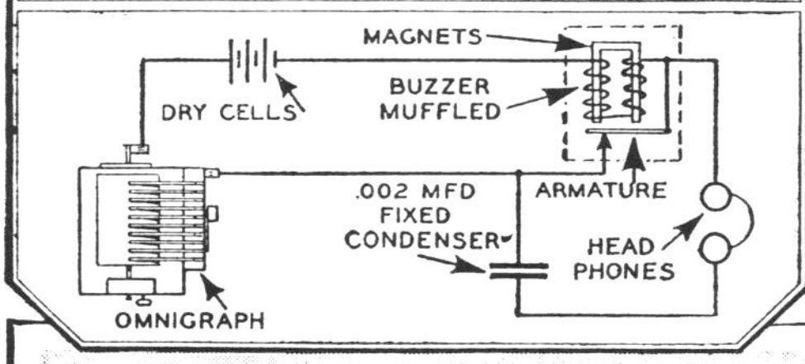


placed on backwards and a new group of signals obtained.

The circuit, shown in Fig. 3, is as simple as could be desired. Taps are taken off the end bearing and switch arm of the omnigraph, the former leading to the battery and the latter going to the armature contact of the buzzer and the fixed condenser.

**Wiring of Omnigraph Is Easy**

Either a loud speaker or a pair of head phones may be used, whichever you have available. In wiring them into the circuit, hook one terminal to the wire which connects the armature to the magnet, and the other terminal to the remaining terminal of the fixed condenser.

Sparking on the springs which make contact with the rotating drum can be prevented by hooking small fixed condensers (about .1 mfd.) across the bearing contact and switch arm lead.

As mentioned before, the speed of the sending depends upon the weight of the lead pieces suspended on the hook. For beginners, about four pounds is required on the hook for transmission at a fairly slow speed.

When practicing, allow the dial to turn around and around till you have learned the letters in that row thoroughly, then switch to the next row. If after long practice you get to know the letters well, shift the switch arm without looking where it is placed and try to take the signals down without a break.

**Practice for Advanced Students**

When you have learned to identify the characters with a moderate degree of speed, try making a few new code sheets, with the letters well mixed, rather than in alphabetical order. With such a scheme you cannot anticipate the letters, and so practice will be much more beneficial.

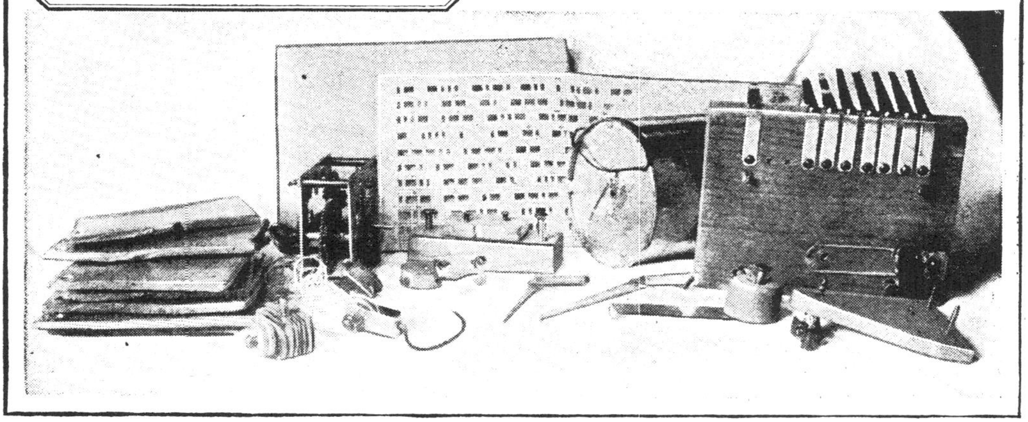

**Fig. 3. Wire up omnigraph and accessory parts according to the circuit at top left. Note that one side of phones taps onto wire of buzzer that connects armature with the magnets. All necessary parts required are seen in bottom photo.**

# "Black Sambo" Sings, Dances to Radio Music

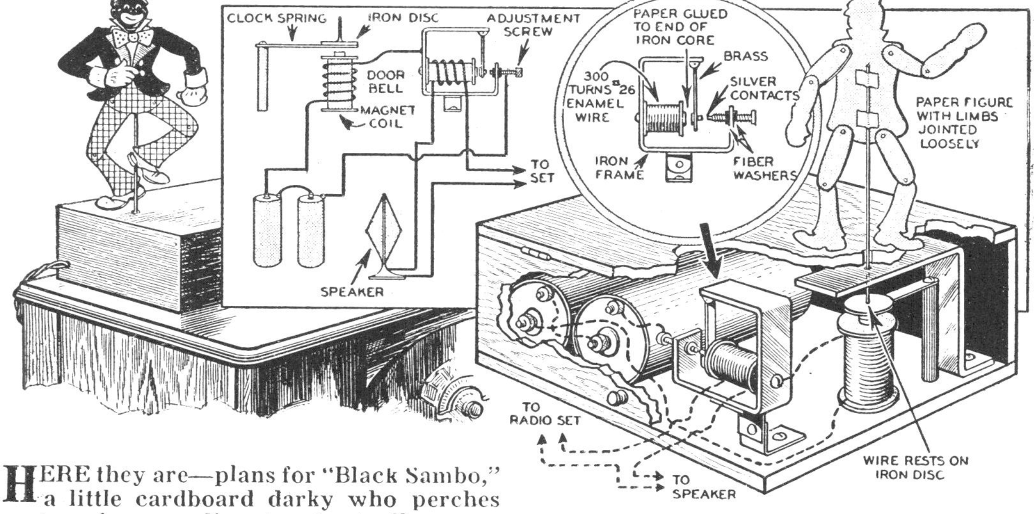


HERE they are—plans for "Black Sambo," a little cardboard darky who perches on top of your radio set and actually moves his lips in time with the speaker or singer at the broadcasting station.

First lay out your design of "Black Sambo" on a piece of stiff cardboard, preferably keeping it not much over four inches in height. Using water colors to paint the face will of course add greatly to the effect.

At the back of the figure the operating mechanism will be mounted. The simple electro-magnetic unit may be made from the coils of an old bell or buzzer, or wound by placing about 300 turns of No. 26 enameled copper wire on a small iron core. This is assuming that the toy is to be used in the output circuit of a high-powered electric set having at least one 250 tube in the output stage. For battery sets and the less powerful electric sets an old telephone receiver having about 1000 ohms resistance may be used, but the moving element will have to be very light.

A groove is cut in a wood base and the cardboard man glued in place. The dimensions of the iron frame and the moving armature are not important, and will depend to a great extent upon the size of the electro-magnet used. A brass bearing for suspending the armature is soldered to the top of the frame. The armature is of brass, with a small piece of iron soldered to the bottom on the side opposite from the core.

Considerable current flows in the loudspeaker circuit of the radio set. This current fluctuates with the volume of sound, varying the pull on the armature.

The toy is connected in series with cone type speakers and in series with the voice coil of dynamic speakers. Some experimenting will have to be done with the armature to get best operation, and a small sliding weight may have to be arranged on the part of the armature to which the lip is attached. The moving part should be as light as possible, and the bearing very free.

### Making "Black Sambo" Dance

This comical dancing darky will jig for hours in step with the radio music, flinging his arms out wide for the peppier rhythms.

The output of a radio set is not great enough to operate this dancing figure, and so a relay and local source of power, such as two dry cells, are needed. A small sensitive relay easily made from scrap parts is connected in series with the radio speaker. This relay in turn operates the stronger electro-magnet which drives the figure.

*(Continued on page 63)*

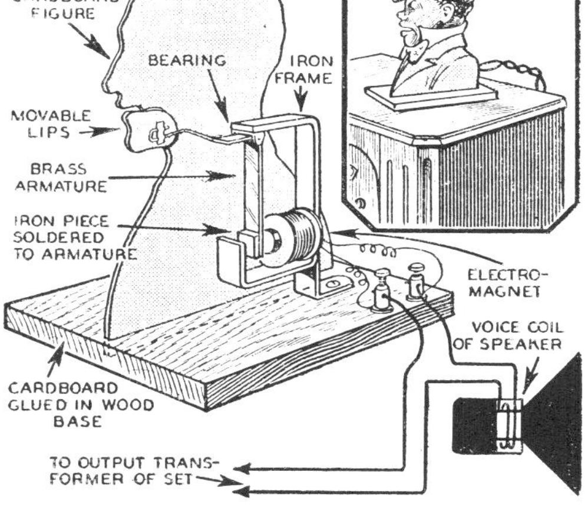

# MAKE PHONOGRAPH RECORDS WITH YOUR RADIO SET

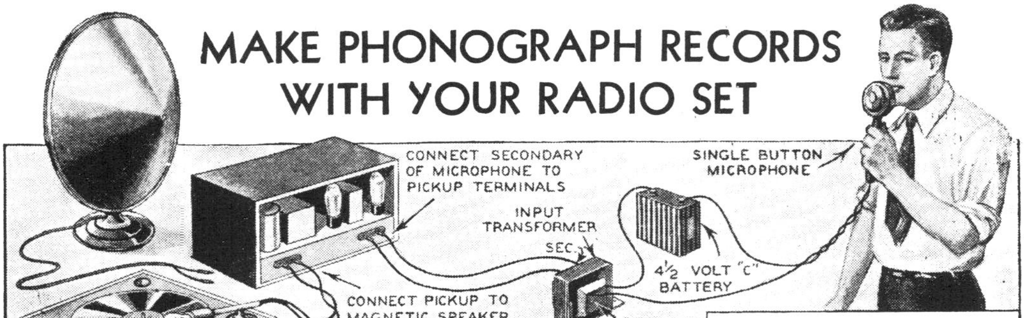


Hook-up for making recordings on radio having magnetic speaker is shown above. Most radios have phonograph terminals.

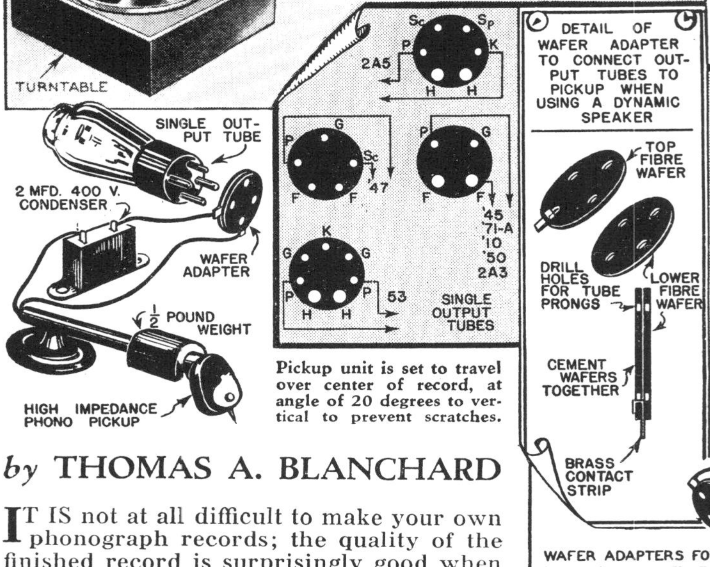


Pickup unit is set to travel over center of record, at angle of 20 degrees to vertical to prevent scratches.

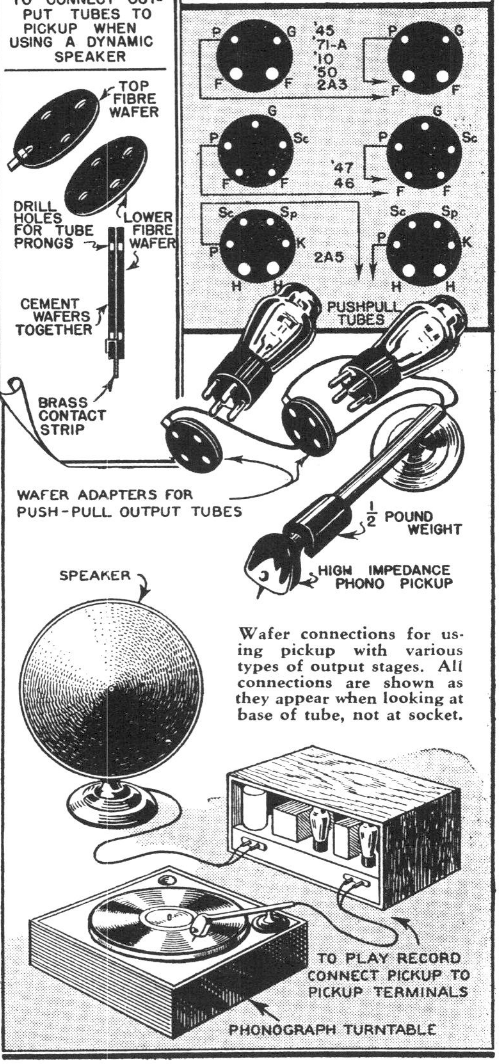


Wafer connections for using pickup with various types of output stages. All connections are shown as they appear when looking at base of tube, not at socket.

Connect apparatus as above when playing back records on magnetic speaker set. Better results are obtained in some cases by connecting pickup first to 4000 ohm output transformer with transformer primary connected to output stage of set.

*by* THOMAS A. BLANCHARD

IT IS not at all difficult to make your own phonograph records; the quality of the finished record is surprisingly good when played back through the loudspeaker of your radio set. You can make recordings of your voice, of musical programs put on by friends, or even of favorite radio programs.

The equipment needed, besides your broadcast receiver, is a good phonograph pickup unit and either an electric or motor-driven turntable.

Pre-grooved records, either aluminum or bakelite, simplify the recording apparatus. A special needle is required for aluminum records, but ordinary needles are used for both recording and playing on bakelite. The bakelite records will be found most economical for home use.

With a radio having a magnetic speaker, the connections are easy. For recording, remove the speaker leads and connect in their place the phonograph pick-up leads.

If you are recording speech or music, connect your single button microphone and a 4½-volt "C" battery in series with the primary of an input transformer, the secondary of which is plugged into the phonograph pickup terminals of the radio set. On radios not having these terminals, connec-

# Electric Phonograph Pick-up Uses HEADPHONE PARTS

IF YOU want a pick-up for the phonograph to the radio amplifier, you can make your own for only a few cents from cast off material about the radio room.

First, get an old stylus arm from a phonograph reproducer. Arrange this so it can be fastened securely to the side of a common headset phone as shown in the detail sketch below.

It will probably necessitate drilling and tapping a small hole in the side of the ear cap for the holding screw of the arm. This should allow the diaphragm pin to come in the exact center of the ear phone diaphragm. Scrape the metal diaphragm clean and solder the end of the stylus to it.

Now make a clamp of thin strip metal that will wrap around the earphone case and leave two horizontals parallel with each other about 1 inch long for bolting to the end of the wooden arm.

The arm should be about 2 inches shorter than a regulation phonograph tone arm, and is hinged as illustrated. When properly arranged, the arm will swing easily up and down on the hinge and turn easily from side to side on the pivot screw.

Wrap the phone wires loosely about the arm and lead them to the primary of a good audio transformer. Wire the transformer secondary to a plug suitable for plugging into the radio amplifier.

By using a suitable needle and good records, very good reproduction should be obtained.

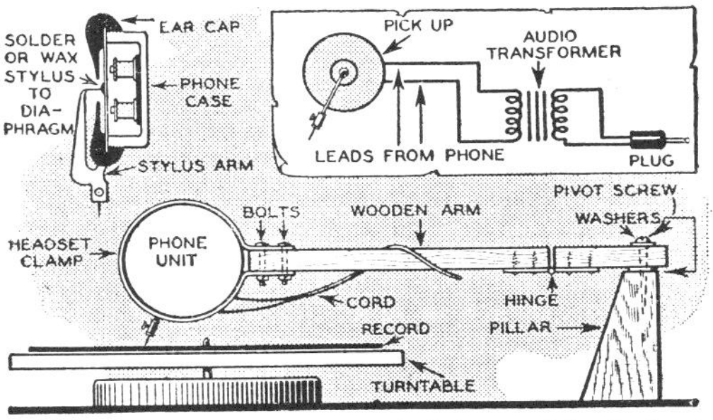


**Ear phone, with stylus arm attached, is seated in the headset clamp, the cord being wound around the wooden arm. The plug inserts in the amplifier circuit of the receiver.**

## "Black Sambo" Sings, Dances

*(Continued from page 61)*

A paper washer is glued to the end of the iron core of the sensitive relay to prevent sticking. A brass armature is used, with a small iron disc soldered to the lower end just under the silver contact, which likewise is soldered in place.

The electro-magnet driving the dancing figure may be made from an old doorbell magnet coil by changing the armature. A strip of spring brass or a piece of an old clock spring, with a small piece of iron soldered to one end, is mounted so the iron comes directly over the top of the magnet core. A metal bracket is mounted over the armature to serve as a guide for the driving wire. A hole a bit larger than the wire used is drilled in this bracket directly over the center of the armature.

## Make Phonograph Records at Home With Your Own Radio Set

tions are made to the plate and cathode prongs of the detector tube with a wafer. With some microphones the input transformer will not be necessary.

The equipment is now ready for use. Turn on the radio and the phonograph motor. Place a half-pound weight on the pick-up unit to make the needle "cut" properly, start it in the groove and begin to speak into the microphone. By watching the record you will know when to stop speaking.

The record can be played back as soon as it is made. Reconnect the speaker and connect the pick-up to the phonograph terminals of the set in place of the output transformer. Remove the weight when playing the record.

It is a little more difficult to connect the recorder to a receiver having a dynamic speaker. In many sets the speaker has a plug-in cable—here it is only necessary to attach the pick-up leads to the voice coil prongs (the two small prongs), using a wafer if desired. Attach to the G and P prongs if you have a 5-prong plug. After making this connection replace the plug in its socket.

A 50,000 ohm potentiometer in series with the pickup will protect the unit from burnouts and provide at the same time a convenient volume control.

Where the dynamic speaker is directly connected to the set, wafers can be slipped under the output tubes to obtain the pick-up connections. Wafers for various types of output stages are shown in the sketches.

In recording radio programs the pick-up unit alone is used. The radio is turned on and the desired station tuned in. At the desired time place the pick-up on the record and begin recording.

# Small Broadcasting Station

**Get "on the air" immediately with this cake tin radio station; it broadcasts through your own radio set from any room in the house with no external connections.**

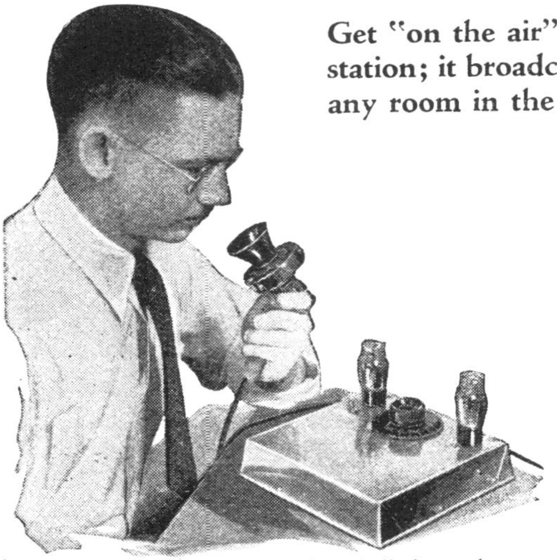

View of completed radio transmitter built into cake pan. Holes are cut in top of pan for tube sockets. Filament switch is in front of left tube. Use large condenser dial. Keep mouth ten in. from transmitter when broadcasting.

MANY of us dream of some day operating gigantic broadcast transmitters—of using powers of 500,000 watts and more to send our voice to listeners all over the world. Some of us may realize this dream —others will not.

Most of us, however, will get just as much pleasure in building and operating this "cake pan" set, the *smallest* broadcast transmitter in the world. It will sound as well over your radio as many of the regular broadcast stations—so well, in fact, that you can fool your audiences in countless ways. Faked police alarms, messages from Mars or from some far off part of the world, or musical programs with home talent are easy to stage with this transmitter.

Broadcasts can originate from any room in the house; two units and two radio-sets make a two way wireless communication system that is equal to the inter-office equipment used by many business firms.

No connections to power lines or to the radio set are needed—just a single ground wire.

### Costs Little to Build

Get "on the air"—a few dollars will buy all of the parts needed.

With only twelve volts of B battery, obtained from flashlight cells, and two cells for the filament supply, this transmitter has broadcast with excellent volume between rooms several floors apart.

All parts are mounted in one end of the pan as shown. This leaves the other end for batteries. Eight batteries in series make up the plate supply, and two in parallel the filament power. The flashlight cells are connected up, bound together in a single unit, and covered with melted sealing wax.

Tubes used are the low drain type —30's. Of course, if the device is not intended to be a portable affair, other tubes such as the

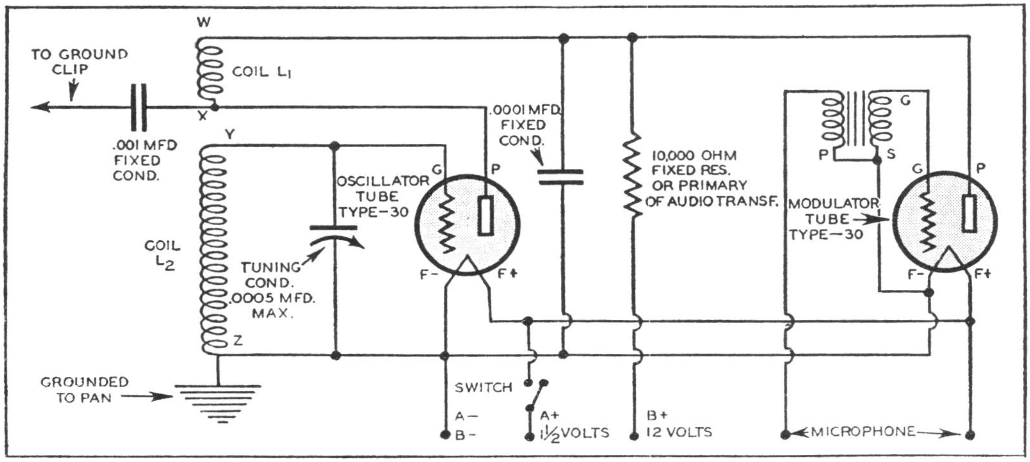


Circuit diagram of the cake tin transmitter. Note that frame of tuning condenser, negative filament leads, and one side of .0001 mfd. condenser are grounded to pan. Parts needed are: 2 four-prong tube sockets, panel type; a 10,000 ohm fixed resistor (or audio transformer); microphone transformer; single button mike; .0005 mfd. tuning condenser; .001 mfd. fixed condenser, .0001 mfd. fixed condenser; filament switch; cake pan; and wire for coil.

# Fits in CAKE Pan

By THOMAS A. BLANCHARD

171, 112, or 201-A types using more filament power can be used.

A regular telephone transmitter, any single button carbon mike, or the microphone described on page 23 of the 1934 Modern Mechanix How-To-Build-It book will be satisfactory. Do not attempt to use a condenser mike.

### Solder All Connections

In wiring the transmitter use heavy wire and solder all connections well. Since the transmitter is in a metal case be sure that none of the wires are accidentally grounded. Lining the inside of the cake tin with insulating paper will prevent trouble.

To use the transmitter set the dial on your receiver to a location where no stations come in. Connect the one transmitter lead to a ground. It will be best to use the transmitter in the same room as the receiver, until you are familiar with tuning procedure.

Next, slowly turn the transmitter dial until a sharp whistle is heard through the broadcast receiver. Carefully tune for greatest volume while talking into the microphone. The transmitter is now tuned to the receiver.

The whistle mentioned above and any accompanying sound distortion can be eliminated by moving the transmitter to the basement or some distant room.

If a telephone transmitter is used, keep about ten inches away from the mouthpiece when you speak.

Due to the extremely low power used it will be impossible for signals to travel any distance. As NO antenna is used NOR should be used, the builder is not violating any federal radio laws by using this transmitter.

Of course, if the current supply is increased and a regular antenna and ground system installed this little device could send a signal possibly 25 miles. *Under no conditions is the builder to use an antenna* unless he has an amateur radio license and uses a coil coming within the assigned "ham" band.

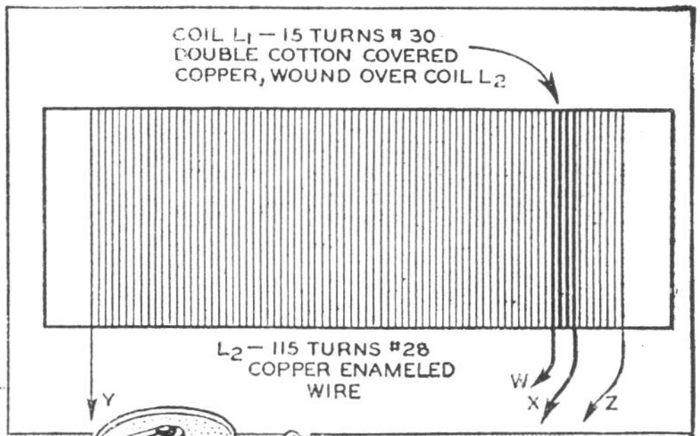


At right: Schematic diagram of coil. Use cardboard or fiber base around 1¼" diameter by 2½" long. Wind coil L-2 first, in single layer with no spacing between turns.

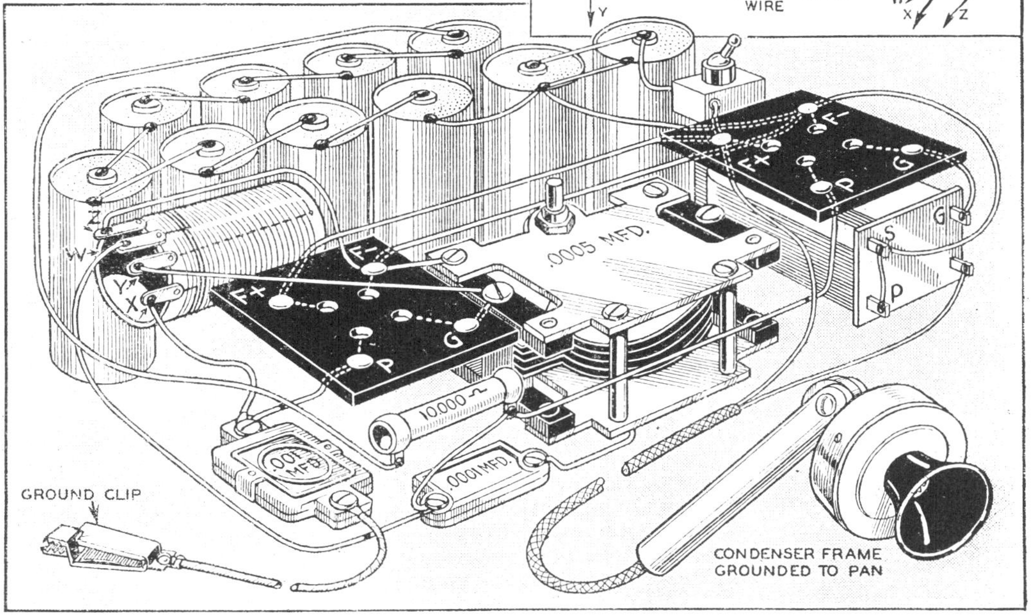


Transmitter as it appears when cake tin is removed. Batteries are on end here to show connections. Condenser shaft, tube sockets, and filament switch are mounted on top of cake tin. Ground and microphone leads run through holes in side.

# This ELECTROMAGNET

Fig. 2. Here is the completed electromagnet, set atop its mounting, with core inserted. With 110-volt current on, it is ready to perform a number of amusing stunts.

Fig. 1. When current is turned on to the electromagnet the bulb glows as if it had a regular connection. The coil under the base is soaked in paraffin to prevent short circuiting.

THE well-known barrel of monkeys could produce no more entertainment than an electromagnetic "circus," consisting of a powerful solenoid magnet and a number of accessories, that you can construct in an evening.

And besides being a source of fun, such a device is highly instructive, and will serve to clear up many of the mysteries of everyday electricity for you.

The electromagnet or solenoid consists of nothing more than a quantity of insulated wire wound on a spool, and provided with a suitable base, connecting wire and plug.

You can obtain a large wood spool from almost any electric shop that does motor repairing; or perhaps the wire you purchase for the magnet will be on a suitable spool. The one illustrated herewith measures approximately 3¾ in. long and 3½ in. across the ends.

**Winding the Electromagnet**

The amount of wire you can use varies within considerable limits. About 3 lbs. of No. 22 enameled, cotton-covered magnet wire will do; or you can use 7 lbs. of No. 20, or 15 lbs. of No. 18 wire.

The larger the wire, the less quickly will the coil overheat. If you have access to a screw-cutting lathe or a coil-winding machine, you can do a neat job, putting the wire on in even layers, with a thickness of oiled cloth tape or other insulating material between layers, as shown in Fig. 6.

You can do an equally satisfactory job by hand. A few inches of each end of the wire should project through holes in one end of the spool.

# Does MYSTIFYING Stunts

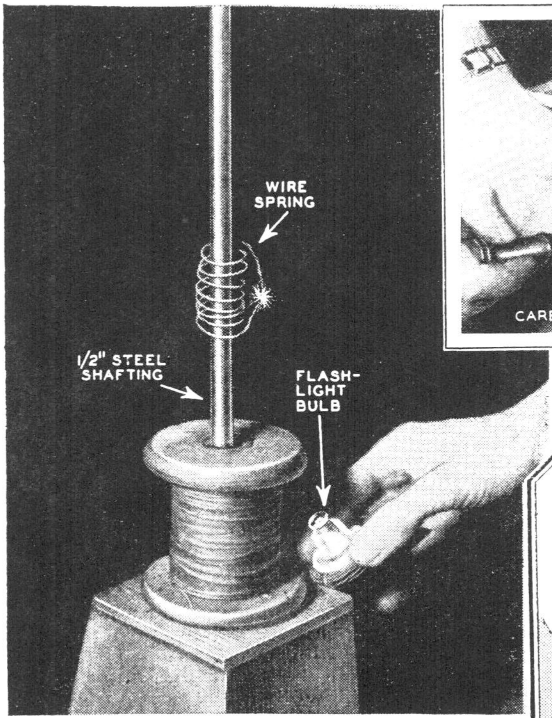


Fig. 3. The wire spring wrapped around the core, which is made of ½-inch steel shafting, dances up and down creating striking display of pyrotechnics. Flashlight bulb wired to coil will glow mysteriously from induced current.

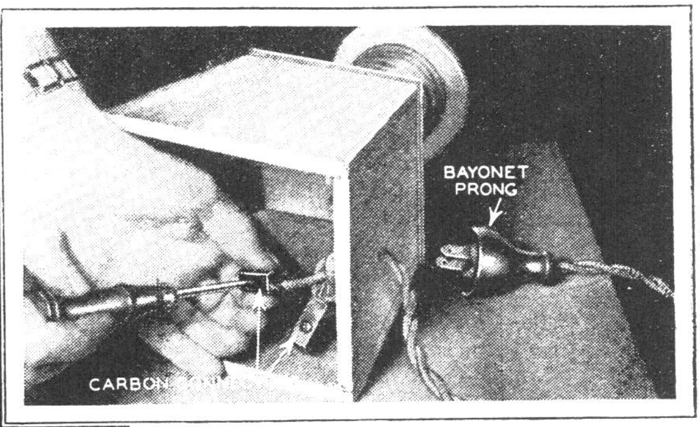


Fig. 4. Electric cord is run through bushing set in box and wired to carbon connectors, to which are also hooked leads from the electromagnet.

Mount the coil in a vertical position, on a hollow wood base so that a core can be moved up and down through the hole in the spool. The base illustrated in Fig. 5 has sloping sides, and measures 5 in. high, 5¼ in. square at the bottom and 4 in. square at the top.

The spool is attached to it by means of two small bolts passing through holes drilled in the end that has the coil wires projecting from it. Also, there is a large hole in the center of the base, corresponding to that in the spool.

### Wiring Up the Magnet

The flexible electric cord to which the coil terminals are attached enters the base through a hole in one side, and is kept from slipping out by a knot tied near the end.

The best core consists of a bundle of soft iron wires held together by a wrapping of cloth, cord or other binder. The core should be of such size that it can be moved up and down in the spool hole, yet will remain in any position.

It should be of such length that, when the lower end is resting on the table or other surface supporting the coil base, the top end will be flush with the upper end of the spool.

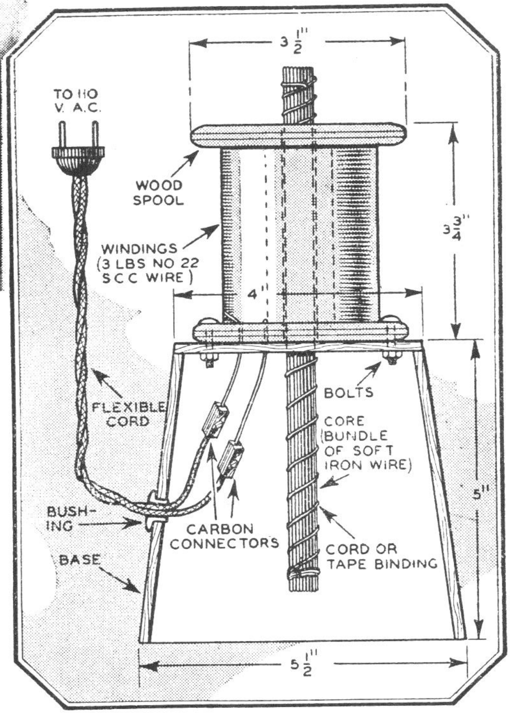


Fig. 5. Drawing shows complete assembly of the magnet and mounting. Spool is secured to box with bolts, while core comes flush with top. Loop cord inside to prevent loosening.

In the model illustrated, this calls for a core 8¾ in. long. Instead of a bundle of iron wire, you can use a length of ½-in. steel shafting, as shown in Figs. 3 and 6, with almost equal results.

You can operate the coil directly from

# *Magnet Makes Ring Fly Off Core—Heats Up Copper Plate Like Stove*

Fig. 6. In photo at left is shown the aluminum ring doing its bounding trick. Center photo shows how wire is wound on spool with lathe. Put thickness of oiled cloth between each layer. At right is seen the copper plate electric stove.

the 110-volt, alternating current house-supply line; or, if you find that it overheats rapidly, you can interpose a resistance in series with one of the leads. An electric heater element, suitably protected by a guard of some kind, will serve. A number of direct-current experiments also can be performed, using a 6-volt storage battery as a source of power.

### Tricks You Can Do With the Coil

Now for some tricks with the coil, using 110-volt A.C. current:

The jumping ring is a spectacular and amusing performer. Adjust the core so that two-thirds of it projects above the coil, or insert the 2-ft. length of shafting into the hole. Drop over it an aluminum ring—a section of ¾-in. aluminum tubing an inch or two long will do, as illustrated in Fig. 6.

Turn on the current. The ring will jump up the core and, if it does not fly clear of the core, will bounce up and down in a swing-like manner, finally coming to rest at a point some distance up the core from the coil.

The height attained depends on the weight. A copper ring, consisting of a single loop of copper wire with the ends connected, will do the same thing, but will not climb as high because of its greater weight.

### The Jumping Ring Trick

Now, with the coil current turned on, grasp the aluminum ring in your fingers and hold it down against the coil end. Soon the metal will become warm, and you may find it necessary to let go of it. The energy with which the magnetic force is trying to push the ring away is converted into heat.

With the core end flush with the coil top, lay over it a thick piece of sheet copper as shown in Fig. 6. Turn on the current, and in a short time the copper will become so hot that water, when dropped on it, will sizzle away into steam. It is even possible to fry an egg on this improvised "stove."

This leads to another interesting stunt. Make a coil by winding, around a bottle or other cylinder an inch in diameter, of 30 to 50 turns of No. 26 insulated magnet wire. Remove the coil from the form, bind it with cord so that it forms a ring, and connect the ends to the terminals of a miniature socket that accommodates a flashlight bulb.

Test the arrangement by bringing the coil near the top of the electromagnet, when the core is all the way down. The lamp should light. If it is too bright, remove some turns from the coil; if too dim, add more.

### Secret of the Flashlight Bulb

With the coil held snugly against the socket, and the bulb in place, dip the wire and base into melted paraffin, covering everything but the glass bulb. Now, if the paraffined coil is placed in a glass tumbler or beaker of water, and the container is set on top of the electromagnet, the lamp will light, in a manner mystifying to the uninitiated.

Another trick involves a dancing coil. Make a "spring" of fairly fine copper or aluminum wire by winding a dozen turns around a form, and arranging the ends so that they touch each other lightly. Drop the coil over the projecting core of the electromagnet, with the current turned on. The coil will dance about in a startling manner, with sparks flying from the ends, if everything has been adjusted properly.

You doubtless will work out many more stunts. For instance, you will find that you have the necessary equipment for making 60 cycle noises when a tin can is set on top of the magnet.

# Converting RADIO Tube Into an Electric EYE

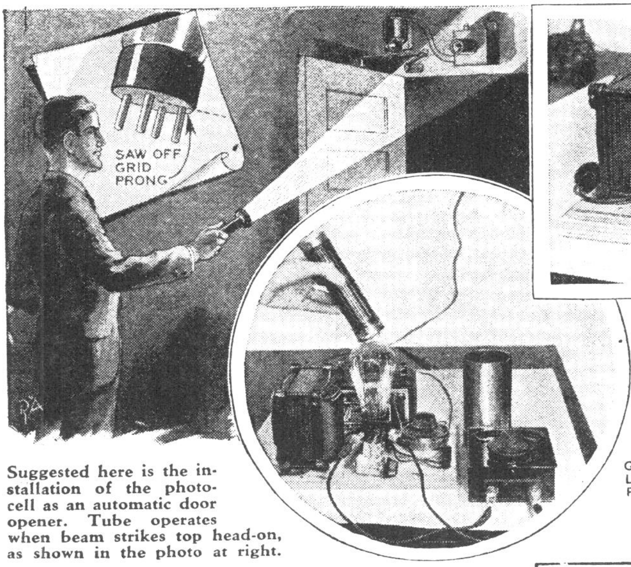


Suggested here is the installation of the photo-cell as an automatic door opener. Tube operates when beam strikes top head-on, as shown in the photo at right.

Here is the electric eye assembled on bread board ready to go into operation. Can covering tube has top cut out to admit light rays.

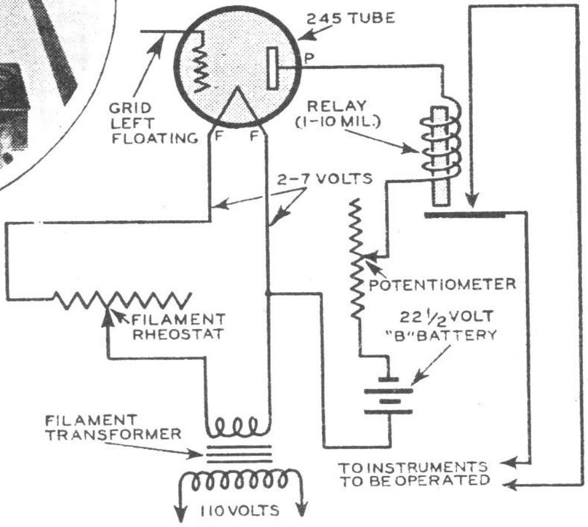


Circuit for electric eye. Note that grid is left unconnected. Dry cells may be used in place of filament transformer. Potentiometer should be about 5,000 ohms. Try different B battery voltages. Milliammeter may be in relay circuit.

THE simple circuit shown here allows ordinary vacuum tubes to be used as photo-electric cells in burglar alarm set-ups, photo-electric light switches, and countless other ways. Though by no means as sensitive as commercial photo-cells, the radio tube will give surprisingly good results with even a low-priced relay.

It has been found that the 210, the 245 and the 250 tubes are photo-electric to a marked degree, and that the light response is sufficient to trip a cheap relay.

Tubes which give the best results are those having open or clear tops so that light from external sources may reach the grid. It is essential also that the grid prong be cut off as illustrated above.

A circuit is given at the top of the page. The filament voltage, you will notice, is regulated by a power rheostat in the secondary of the transformer circuit. The experimenter will have to discover the best filament voltage by tests.

The relays used in connection with these new photo-electric cells should be capable of operating on a range between 1 to 10 milliamperes. RADIO BUILDERS' MANUAL editors will be glad to give you the addresses of manufacturers of relays.

When operating the tube, remember that the more light that reaches the grid of the tube, the greater will be the current output. Therefore it is advisable that the tube be installed in such a position that the beam strikes the top head-on, as illustrated in an accompanying photo.

## Filter Short Wave Interference

ON MANY modern midget superheterodyne receivers it is very difficult to eliminate signals from nearby short wave transmitters operating on or near the intermediate frequency of the set. It is quite common to pick up these naval, airport, or commercial stations regardless of where the set is tuned on the broadcast band.

The easiest way of eliminating this annoying disturbance is to place a tuned filter in series with the antenna lead as shown in the sketch at the left.

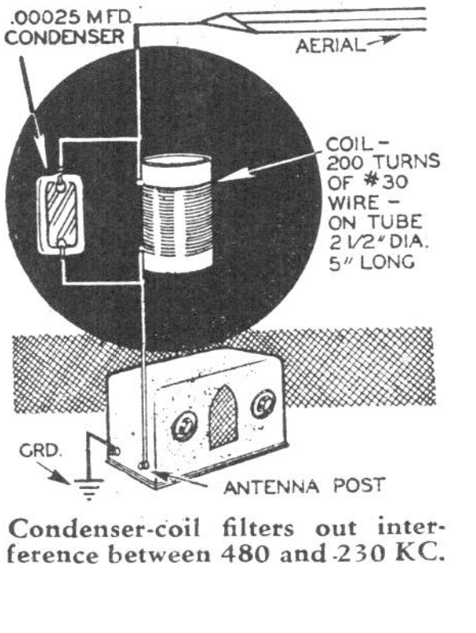


Condenser-coil filters out interference between 480 and 230 KC.

# Talk on 5-METER Waves

**Hold phone conversations with friends on lakes, at camp, or in the city, as far as ten miles away, with this compact but powerful ultra-short wave combined transmitter and receiver.**

ONLY a few years ago radio transmission on the five meter band was a rare thing. Today ultra-short wave sets are giving hundreds of fans a chance to operate their own transmitters and enjoy the thrills of amateur radio without putting out hundreds of dollars for equipment.

Designed especially for portable use on autos or boats, this 5-meter combination transmitter and receiver can be built from standard parts for about $5. Many fans will be able to salvage parts from old battery receivers. The tubes, batteries, headphone set and carbon microphone complete the setup.

Actually we have here a private telephone system, with a range up to ten miles. With one set at home or camp, and another mounted for portable use, telephonic conversations can be carried on from your car, boat, or even while hiking.

For portable use the set may be mounted in a metal tool box or overnight bag. Special small batteries should then be used to cut down weight.

The transmitter-receiver is mounted entirely on an aluminum panel 5¼" x 5", which also serves as chassis and ground. No ground wire is used.

The arrangement of parts is novel, but radio experimenters will recognize the transmitting unit as a split Colpitts circuit modulated by a '33 pentode. With the switch thrown for receiving, the '33 tube becomes a stage of audio frequency and the transmitter a receiver operating very much like a super-regenerative set.

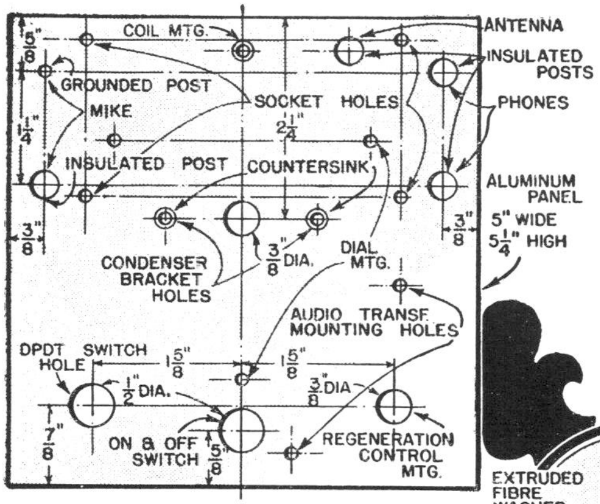


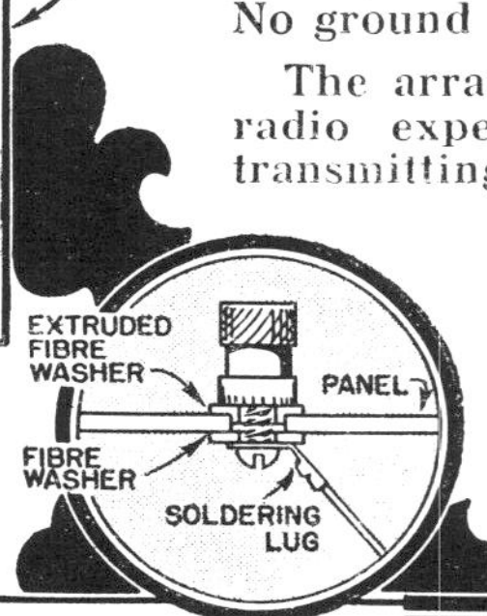


Lay out and drill ⅛" thick aluminum panel as shown above. All parts are mounted directly on this panel or set off from it with brackets and studs. Audio transformer mounts on two studs, condenser insulating panel on two, coil on one. At right: Use of fiber washers as insulators.

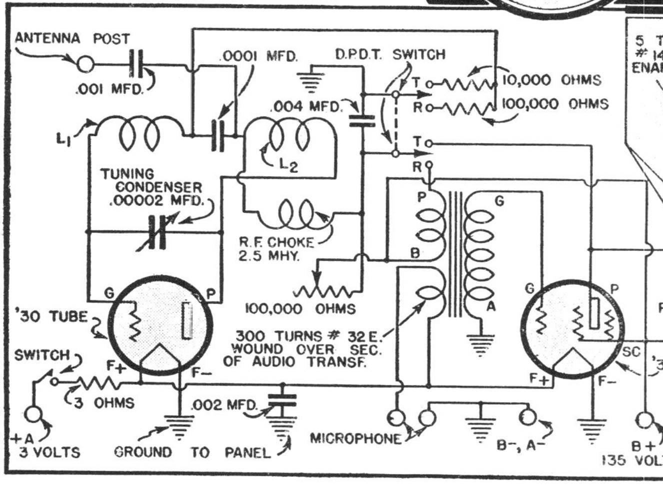


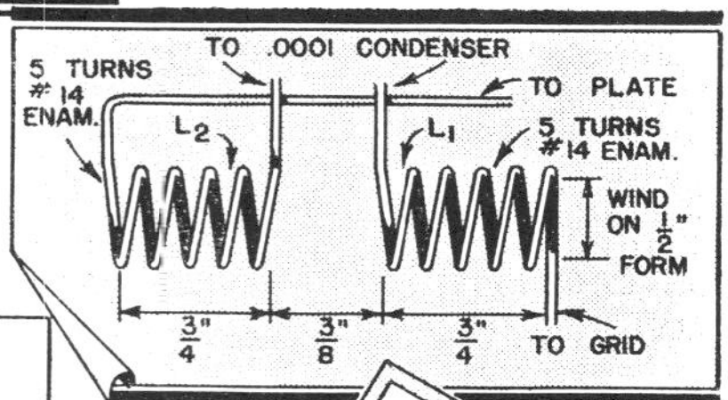


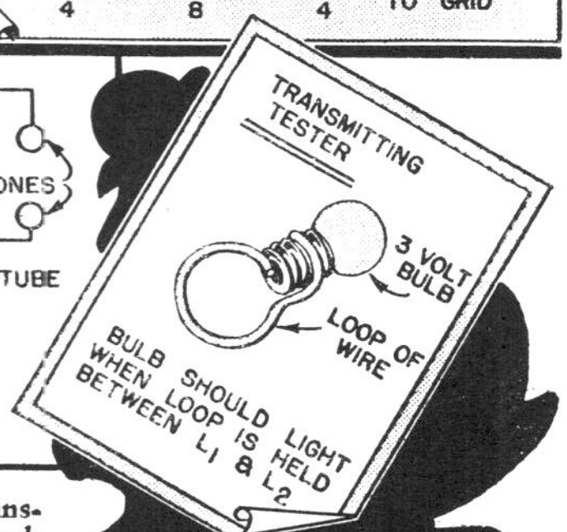


Circuit diagram of M-M Transceiver. Remove laminations of standard audio transformer, wind on extra turns for microphone input as indicated, then re-pack laminations. Details of coupling coils and transmitter tester are shown at right.

# with M-M TRANSCEIVER

By
THOMAS A. BLANCHARD

Microphone and audio transformers have been combined to save space by winding between 250 and 300 turns of No. 32 enameled copper wire over the secondary winding of an ordinary audio frequency transformer. If desired, a manufactured transformer serving exactly the same purpose can be had for around one dollar.

The transformer and tube sockets are mounted on metal or fiber studs to eliminate body capacity. The small tuning condenser is mounted on a strip of bakelite which in turn is held away from the panel by studs. This condenser should be no larger than 20 mmf. (.00002) in order that the set will tune down low enough; this size generally will have seven plates. Sockets are of the wafer type, mounted as shown, with the coils located between.

The coils are wound with No. 14 or No. 12 plain or enameled copper wire. Both coils have 5 turns wound on a ½" dowel or pipe form. When removed from the form the coils will open slightly, giving the correct finished inside diameter of about ⅝". Spread out the windings to leave about ⅛" space between each turn.

*(Continued on page 72)*

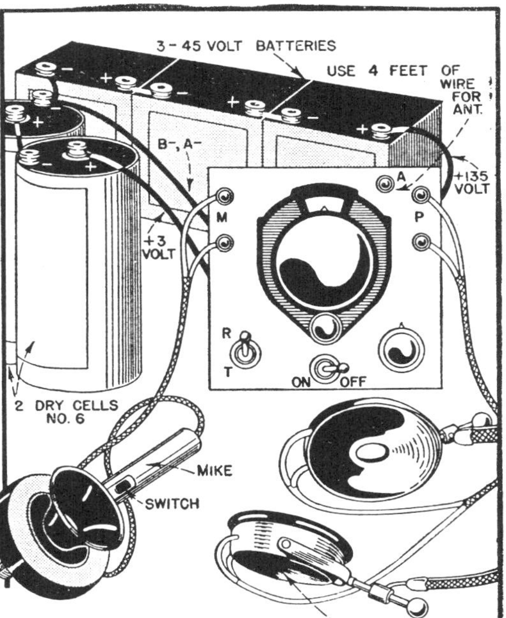


Batteries, phones, microphone are easily connected to transceiver if above scheme is followed. Use any single button carbon hand microphone having a built-in switch.

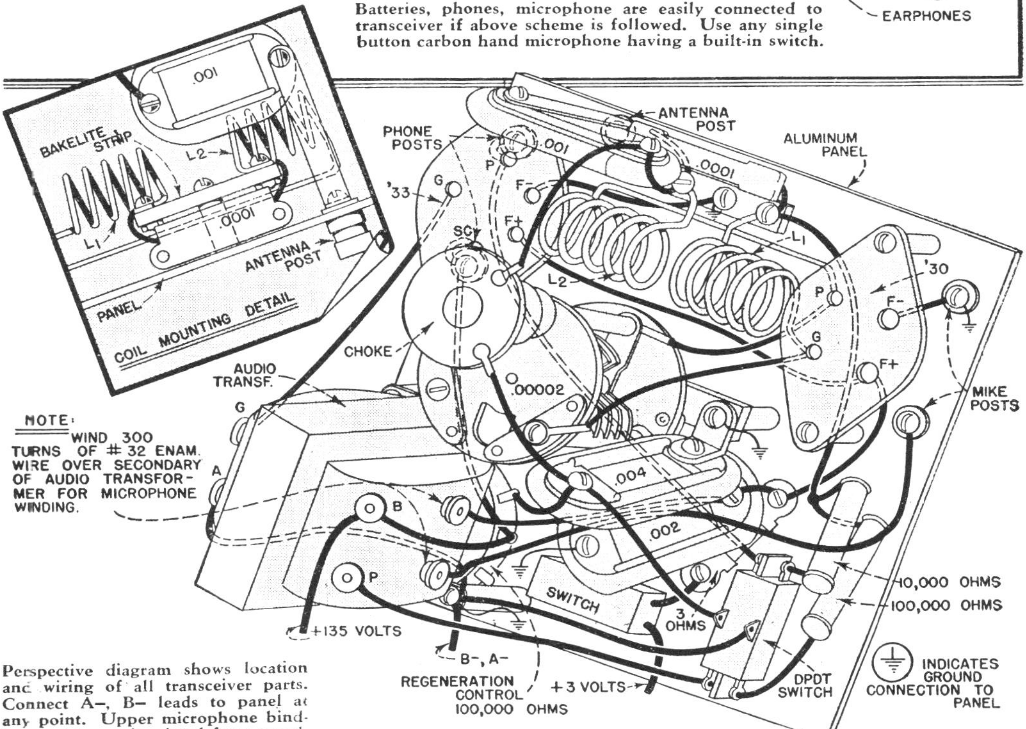


Perspective diagram shows location and wiring of all transceiver parts. Connect A−, B− leads to panel at any point. Upper microphone binding post is not insulated from panel.

# Convert Your Radio Headphones Into Extra Loudspeaker

ARE you shy a speaker for emergency use? Well, if you have an old headphone that is gathering dust, here is a speaker that will fill the bill for amateur transmitting use or for use in camp.

Get a piece of board for a base and clamp one headphone to the center near one end. At the opposite end mount a block about twice the height of the phone.

Next find a nice clear-grained shingle and fasten a piece of lattice or smooth lath to the butt end as shown and hinge the opposite end to the block of wood.

Lower this stick over the phone and mark a point directly over the center of the phone diaphragm. Drill a small hole there and drive in a pin or piece of stiff wire sharpened at the bottom end.

### Wire Rests in Wax on Diaphragm

Arrange the wire so when the point touches the diaphragm the stick will be about horizontal.

Lift the stick, place a small bit of wax on the diaphragm and lower the point into it again. Allow to harden and your speaker is ready. Connect the headphone cord to the receiver output and you will be surprised at the volume and quality from this speaker when a good station is operating.

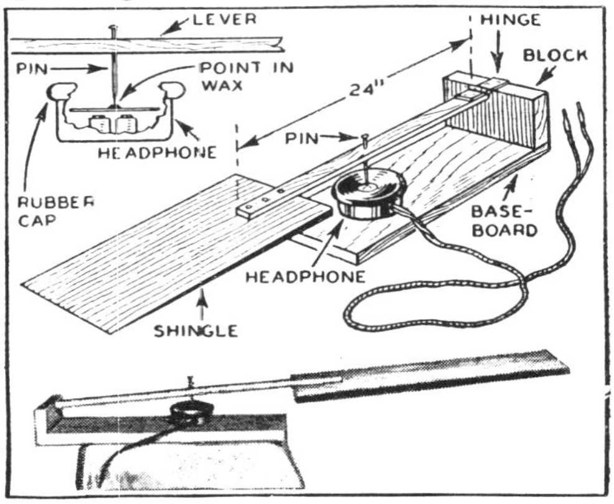


Pin or wire, driven through lever, rests on lump of wax on the diaphragm of the headphone, which is wired to the loud speaker terminals of your receiver.

## Talk Over Five Meter Waves With the MM Transceiver

*(Continued from page 71)*

A small strip of bakelite mounted on a stud was found to be the simplest method of mounting the two coils. Leads are fastened underneath to two small bolts, while the inner coil terminals are soldered to the heads of the bolts. A midget .0001 mfd. fixed condenser connected directly across this mounting aids in maintaining rigidity.

The 100,000 ohm potentiometer used as a regeneration control is mounted in front of the transformer. A separate single pole single throw toggle switch is used in the filament circuit, and a double pole double throw switch to change over from sending to receiving. Unless the microphone used has a switch mounted directly on it, it will be necessary to remove one lead from the panel binding post when receiving.

The seven plate midget tuning condenser is held off from the panel by a bakelite strip mounted on two studs. A bakelite or hard rubber shaft extension must be used to connect condenser to vernier tuning dial.

### Only 4-foot Antenna Needed

All leads which are shown grounded in the circuit diagram are connected directly to the aluminum panel to cut down on wiring.

A very short antenna is required for ultra short wave transmission. A four-foot length of copper wire attached to the antenna binding post and insulated from nearby objects will work satisfactorily.

The transmitter will operate on 90 volts, the voltage required for the receiver, but far better results are to be had if three B batteries, giving 135 volts, are used.

Filament current can be obtained from a 2-volt storage battery—in this case the 3-ohm filament resistor is omitted. If two dry cells are used the 3-ohm filament resistor must be used to cut down the voltage to 2 volts.

To put the set in operation, throw the switch to receiving position, close the filament switch, and turn the regeneration control slowly until a loud sh-sh-sh is heard. After locating the proper spot on the dial for reception of a nearby 5-meter transmitter, the dial setting is left alone for both sending and receiving.

Recent changes in amateur radio license laws permit portable operation on the ultra-short waves without usual notifications to the radio inspector. It is still necessary, however, to have an amateur radio operator's license to legally operate this five-meter transmitter.

**M-M TRANSCEIVER BLUEPRINTS**
For those readers who prefer to work from large-size shop drawings, a set of blue-line prints on this project have been prepared to sell for 50 cents postpaid. Included with the prints is a detailed list of the parts required. Copies of the booklet "How to Become a Radio Amateur" are available at 25c postpaid. Send your order to Modern Mechanix Publishing Co., 529 South 7th St., Minneapolis, Minn.

# Telephone System Uses RADIO Headphones

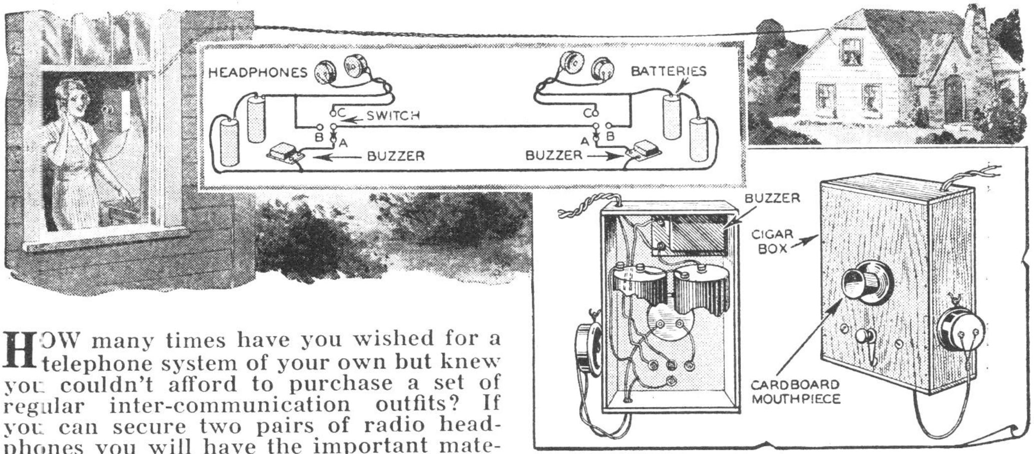


HOW many times have you wished for a telephone system of your own but knew you couldn't afford to purchase a set of regular inter-communication outfits? If you can secure two pairs of radio headphones you will have the important material needed to construct a fine little telephone outfit. With a few other parts such as two 20c door buzzers, two cigar boxes, homemade tap switches and several dry cells you can put together a very efficient little phone system that will work up to several thousand feet.

First of all remove the headbands from the phones. These can be snapped off very easily. Next, get two suitable cigar boxes. In the front of each drill four small holes for the homemade tap switch, and a large hole in the center for the phone unit.

Mount the buzzer inside of the box, allowing room for two dry cells. Connect up the tap switches exactly as shown. These are made with a strip of brass for a tap contact and small brass bolts for taps.

The second box is connected in the same manner as the first. A screw eye is fastened to the side of each box to serve as a hook for the phone that will be used as receiver.

Assuming that the outfit is completed and we are ready to make a call we proceed as follows: Both tap switches are set on "A". To call the party at the other end place tap contact on "B" and after a few seconds push up to "C". Answering party changes his tap switch to "C," and begins conversation.

In the event that there is no answer return the contact to "B" for a few seconds more. This is the tap which calls the party (sounds buzzer). After the conversation is finished both parties return the contact to "A."—Thomas A. Blanchard.

## Improvise a Variable Grid Leak From a Discarded Filament Rheostat

FIXED grid leaks have gradually come to supplant the variable style, but for the experimenter the latter type is very handy. A good one can be made from an old rheostat and some firm cardboard as shown in the sketch.

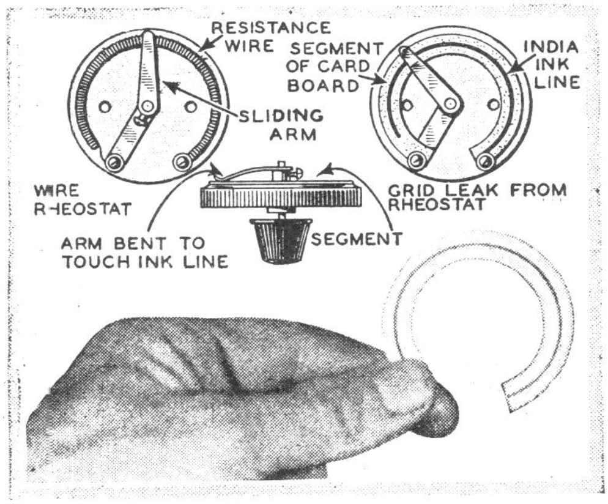


Lever makes contact with India ink line to vary resistance.

First remove the resistance wire from the groove in the rheostat shell and take off the sliding arm. Then, with a compass, draw a pair of circles on a piece of hard-surfaced cardboard.

Cut this out to form a sort of horseshoe and lay over the groove in the shell, fastening it down with the two screws at the bottom.

Now draw an arc along the center of the cardboard with India ink (drawing ink), starting an eighth inch from the screw attached to the sliding arm bridge and ending at the opposite screw so to make contact with it.

Polish the bottom of the sliding arm to a glass-like finish and replace the arm. Adjust it so it makes light but firm pressure on the inked arm.

By sliding the arm along the line the resistance can be quickly varied. With a line about 1/32-inch wide, the resistance will run into 6 to 8 megohms.

The resistance can be varied from 0 to 1,000,000 ohms simply by twisting the knob.

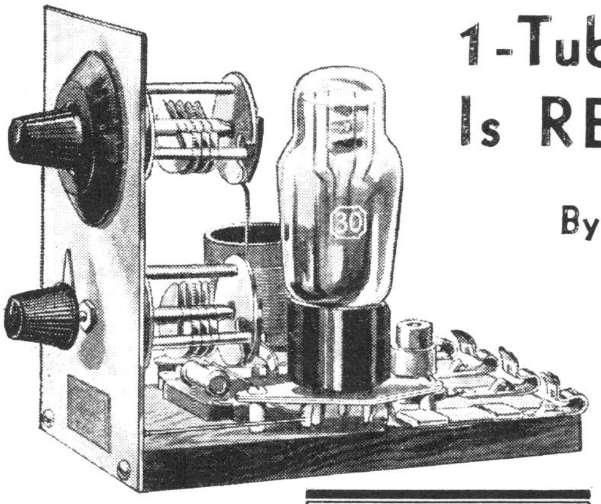

# 1-Tube Short Wave Marvel Is RECEIVER or Adapter

By THOMAS A. BLANCHARD

Completed short wave set weighs only one pound. It is very important that coils be wound exactly in direction shown, and connected properly. D, E, and F bring in most stations.

| PLUG-IN COIL DATA | | | |
|---|---|---|---|
| COIL | METERS | NO. 1 | NO. 2 |
| A | 18—25 | 7 | 7 |
| B | 25—35 | 10 | 10 |
| C | 35—45 | 14 | 15 |
| D | 45—65 | 18 | 20 |
| E | 65—100 | 50 | 50 |
| F | POLICE | 100 | 100 |

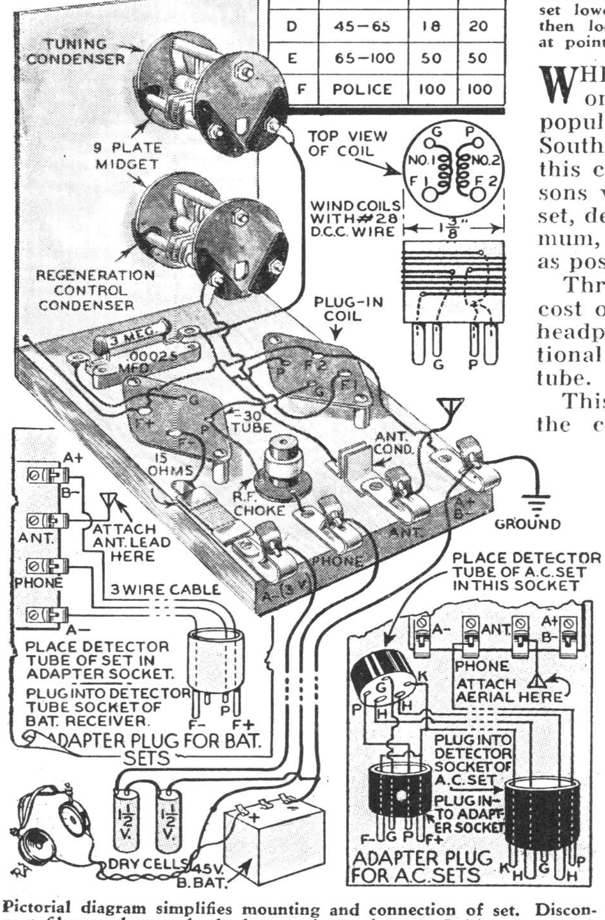


Pictorial diagram simplifies mounting and connection of set. Disconnect filament battery lead when set is not in use. Below are connections enabling receiver to be used as adapter on either all-electric or battery receivers. Ground of regular receiver is unchanged for short waves. Short antenna condenser leaves to get broadcast band.

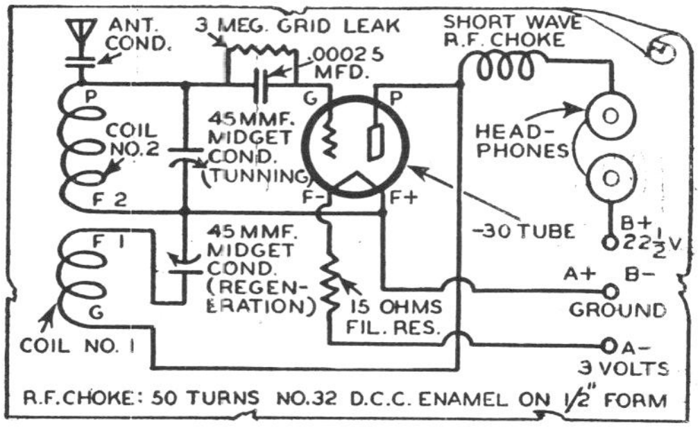


Circuit diagram of one tube short wave marvel. In tuning, set lower dial at point where set just stops squealing, then locate station with top dial. Reception is loudest at point where set just stops regenerating or squealing.

WHETHER used as a short wave receiver or adapter, this little set will bring in popular short wave broadcasts in Europe, South America, and Canada as well as from this country. Designed primarily for persons who have never before built a radio set, details have been kept down to a minimum, and construction simplified as much as possible.

Three dollars will just about cover the cost of parts. Most everyone has a set of headphones on hand, so two dollars additional will take care of the battery and tube.

This set uses a metal panel to which the condenser rotors are automatically grounded, consequently there will be no annoying "squawks" each time you touch the dial.

Make the metal panel and the wood baseboard each 5x3½" in size. The parts are mounted on the base with wood screws. Wafer type sockets, requiring much less space, are used. The gridleak and grid condenser, of the midget type, are mounted directly under the regeneration condenser.

The radio frequency choke is the latest low-loss type consisting of 50 turns of No. 32 double cotton covered or enameled wire wound on a ½" bakelite form. You can make your own choke by winding the same number of turns of wire from an old audio transformer on a ½" diameter wood, bakelite or glass form. The 15 ohm filament resistor reduces the dry cell

# A Simple SIX-VOLT Storage Battery Charger

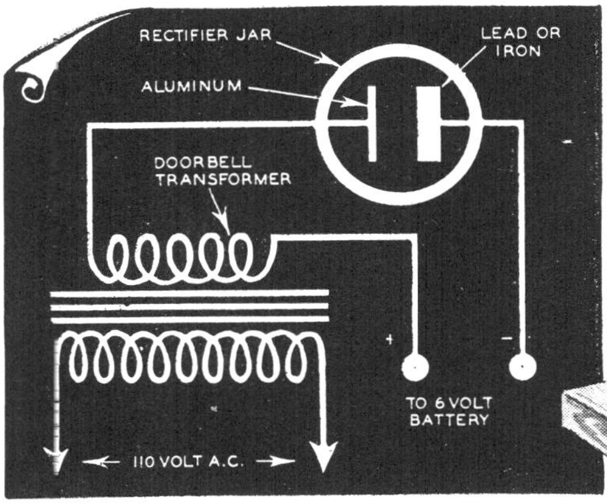


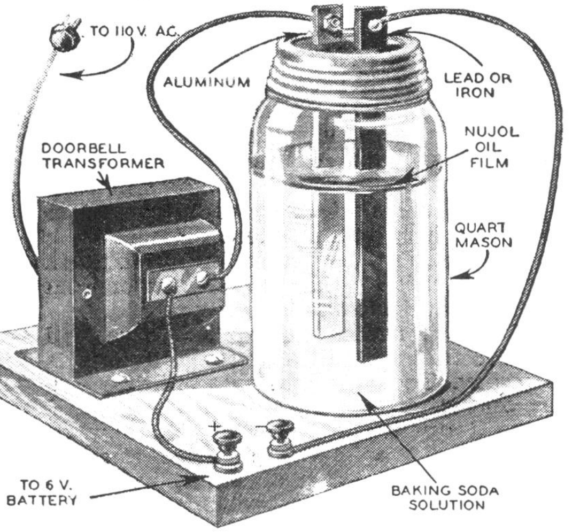


Chemical rectifiers must be formed initially; reforming is necessary whenever unit is allowed to remain idle for any length of tme. Connect cell across high voltage secondary of radio power pack transformer having 400 to 600 volt output. Be sure to place parallel bank of lamps in series with 110 volt primary of transformer, since unformed cell acts as dead short for few minutes, and may blow fuses or burn out transformer. As rectifier begins to form, use larger lamps or more of them in parallel until rectifier cell withstands full voltage. It can now be used in charger circuit. For best results solution and electrodes should be as nearly chemically pure as possible. Size of plates determines current which will pass.

HERE is the simplest possible storage battery charger. It can be connected to your car or radio battery over night, and in most cases the cells will be fully charged by morning.

Take a quart Mason jar, cut a large hole in the zinc cover, and insert a fiber or wood insulating disc. Drill holes in the disc to take the electrodes, one being of thick aluminum and the other of iron, copper, or lead. Electrodes are 1"x6", and mounted about one inch apart.

Fill the jar with a dilute solution of sodium bicarbonate (baking soda), using ½ oz. to a gallon of water. Borax may also be used; here make a saturated solution, but be sure no borax settles to the bottom of the jar. Pour a film of clean oil of any kind over the solution to prevent evaporation.

Attach terminal screws to the outer ends of each electrode, and connect as indicated.

## One Tube Short Wave Marvel Is Receiver or Adapter

voltage to two volts for the type 30 tube filament.

The antenna condenser is nothing more than two strips of phosphor bronze mounted directly on the wood base. The upright portions are ½" square and 1/64" apart.

Fahenstock clips are used as terminals. The ground, A plus and B minus leads go to the same clip, and one phone terminal goes directly to the plus terminal of the B battery. Any set of earphones will work; if you buy a new set, get a pair with a resistance of 3,000 ohms.

### Using Set as Adapter

Short wave reception with loudspeaker volume on your regular broadcast receiver is possible if an adapter plug is used.

The adapter plug for battery receivers is very simple. Attach a 3-wire cable to a four prong tube base as shown in the sketches, and run these leads to the terminal clips of the short wave receiver. Place the detector tube of your radio set in the tube socket of the short wave receiver (now an adapter), and insert the four prong plug in the detector tube socket of the receiver. Attach the antenna lead to the terminal on the adapter, leave the set ground as it is, and tune in. Set the broadcast receiver on a band at which no stations are received, and tune with the adapter controls.

The adapter plug for an electric receiver is somewhat more difficult to hook up. A button type 5 prong socket is mounted inside a 4 prong tube base, and a hole drilled in the side of the tube base to bring out leads to the adapter plug. A 5 prong adapter plug is used, this going into the detector tube socket of the broadcast receiver. The detector tube plugs into the tube base socket, and that in turn plugs into the tube socket of the short wave adapter. If bus bar wire is used to connect up the socket to the 4 prong tube base, and the wires pulled down through the tube prongs, it will not be necessary to cement the socket in place. Tuning is exactly the same as for the battery set adapter.

# Antennas and Grounds for

THE problem of obtaining antennas and grounds for a radio set installed on a boat often involves difficulties in the way of getting necessary antenna length and a good contact with the water. There is no need to leave your radio at home when going for a cruise. A good radio is just as much company on the water as it is on land; perhaps more so. So why not begin to think about equipping your sailboat, motor boat or even canoe with a receiving set?

Nowadays radios are so sensitive and selective that very little antenna system is necessary. Too, sets are being built for battery operation that are single dial control and will operate off the auto battery or even from dry cells with good volume. So let's see what we can do about putting a radio in the old boat and getting even more out of the summer's cruise than usual.

**Water Contact Makes Ideal Ground**

First comes the subject of grounds. Well, that's easy. A ground actually means contact with the earth and the latter includes land or water. Your boat sits in water so what could be easier?

Any large metal surface continually below the water-line will do the trick. For the large motor boat or sailing craft an excellent scheme is to fasten a sheet of copper to the bilge or bottom and lead it right into the cabin to the ground post of the set.

The drawing below shows how this can be done. Cut a strip of sheet copper in a short stemmed T. Make it a half to a third the length of the boat. Fasten the horizontal portion to the hull with copper tacks and lead the stem up to the moulding under the gunwale.

**Ground From Metal Plate on Hull**

Knock out some of the caulking between two of the planks and, bending the stem at right angles, push it inside, to end in a Fahnestock clip or binding post for the ground lead of the receiver. Then recaulk the seam and paint the copper the color of the hull and it will not be noticeable.

Of course if the boat has a steel hull then there is nothing to do but fasten the ground lead directly to it in some convenient place.

For rowboats, skiffs and canoes an ideal way of obtaining a ground is to turn the boat over and fasten a piece of stiff copper wire to the center of the keel with copper staples, as illustrated on the opposite page. Place them a few inches apart to hold it

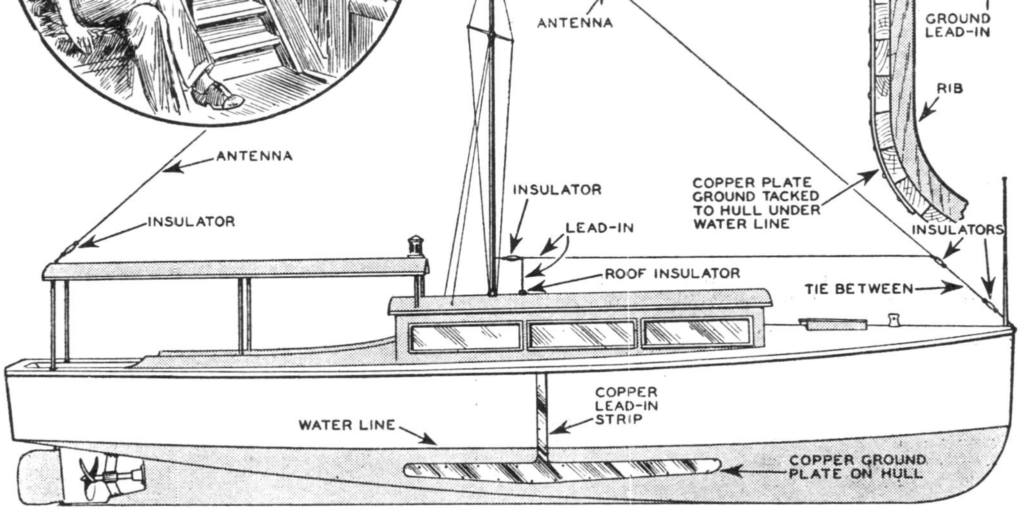


**Antenna for medium-sized power boat is strung up with insulators at points shown in this drawing. An inverted T-shaped piece of copper, tacked to the hull to make contact with the water, forms an ideal ground for operating broadcast receiver.**

# Your Canoe or Boat Radio

**Here is best method for stringing up antenna on your canoe. For a ground, fasten length of wire to the keel with staples. Half round of bow will form satisfactory lead to ground post on the receiver. Method works well on all small boats.**

snug and lead one end, in the case of a row-boat, up the bow or stern and inboard where it can be connected to the set.

A canoe usually has a metal half round extending to the top of the stem from the keel so it is only necessary to solder each end of the keel wire to this and then connect the set to the top end of this metal on either bow or stern.

Now let's think about antenna systems. For a sail or power boat with a mast the thing is greatly simplified.

### Antenna for Canoes, Skiffs

For canoes and small boats you can use a bamboo fishing pole mast stepped into the bow with an insulator at the tip. Run the lead-in and antenna from that to the set and to a second insulator at the stern.

An excellent antenna that takes up no overhead space can be obtained by trailing about 50 ft. of rubber covered wire from the stern and in the water. The outer end of the wire must be sealed in sealing wax and rubber-taped to prevent water from coming in contact with the wire inside. This of course works best when the boat is in motion. You'll have to haul in wire when you stop.

No matter what antenna system you use always provide a lightning arrester just as you would for the house. Insurance laws on larger boats require this.

Now let's consider power supplies. For the filaments you can use the starting or lighting battery if it be of the 6 volt storage variety. Just clip the filament circuit to it. The auto type 6 volt tubes are admirably adapted to a marine radio and will work just as nice as in a car.

If the boat carries a 12 or 24 volt battery then it will be necessary to cut down the voltage with a power rheostat to reduce it to the required filament voltage. If your motor runs on a magneto only, then you had better use the new 2 volt tubes or 199's and use dry cells for power.

Plate power should be from a bank of B batteries. Three 45 volt blocks of B battery will suffice for any ordinary receiver.

With the advent of auto radios has come the spark suppressor. This is attached between the spark plug and the cable and will effectively cut out all sparking noises in the receiver so that you can cruise about with the engine going and not experience any difficulty about hearing programs.

Short wave receivers operate particularly nice on board boat because even the simplest requires only a very short antenna.

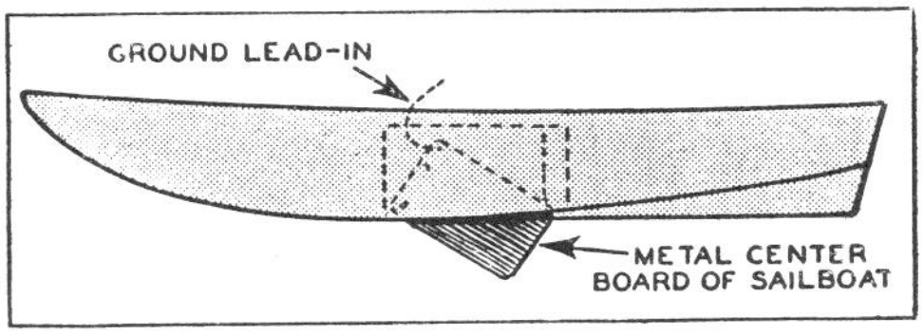


**Metal center board of sail boat, making contact with water, acts as efficient ground for your sea-going broadcast receiver.**

# POLICE CALL Box Brings

*by* THORNTON HALLET

**Down on the short waves police radio stations are constantly transmitting orders to squad cars cruising about the city on the lookout for lawbreakers. With this little call-box hooked to your broadcast set you can tune in on these signals and thus have a virtual ringside seat at battles waged by police against criminals.**

The Police Call-box set up for an exciting evening at listening in to calls from police headquarters to squad cars.

"CRUISER number 16. Go to Corner of Chestnut and Main Street. Suspicious man lurking about.—Cruiser number 4. Proceed cautiously up Empire Road and follow Taxi number —. Escaped convict said to be passenger."

Do you get this kick from your broadcast receiver? If you don't then you better build one of these Police Signal Call-boxes and get all the thrills of a hair raising movie. from real life.

The construction is very simple. First. you will need a panel of bakelite, hard rubber or thin dry wood. And as this call-box was built only from what could be found about the shop, any definite measurements or specifications would be useless. Circuit constants however should be about as given.

Dimensions best to follow are given in Figs. 3 and 4. Drill panel holes about as indicated so your particular condensers, socket, etc., can be arranged somewhat as indicated in Figs. 3 and 4.

Following the schematic diagram you can then wire up in connection with the picture diagram (Fig. 2) which should

Here's how your call-box will look when completed. The flexible cord leads to the base in the receiving set.

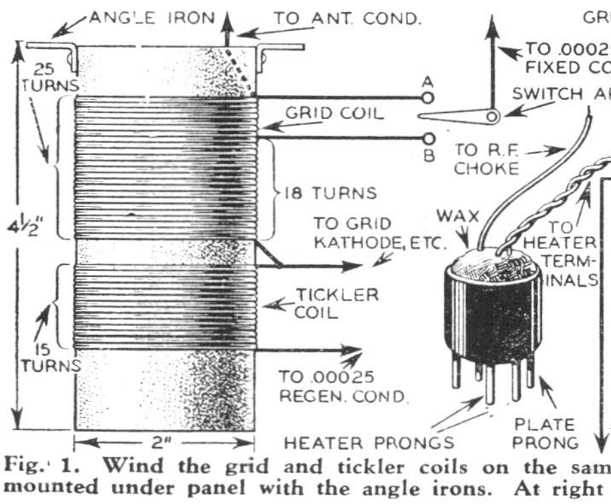


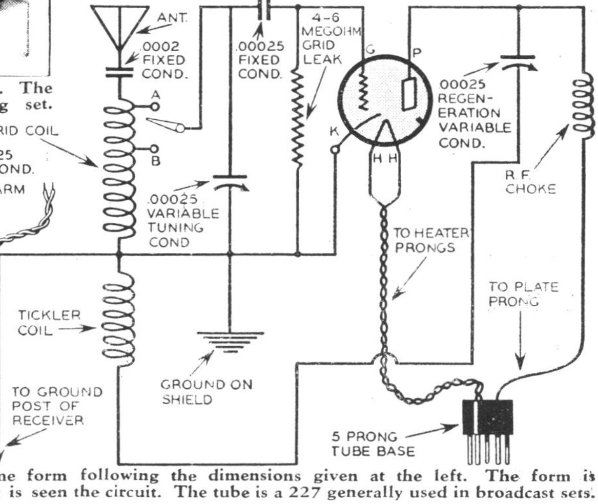


Fig. 1. Wind the grid and tickler coils on the same form following the dimensions given at the left. The form is mounted under panel with the angle irons. At right is seen the circuit. The tube is a 227 generally used in broadcast sets.

# You SQUAD CAR Orders

make it very easy for anyone to hook up without error.

What to do about the grid and tickler coils? For this rig use a 2 inch form about 4 inches long. The grid coil consists of 25 turns of No. 22 DCC wire wound close together. Tap off at the 7th turn from top end. The tickler consists of 15 turns of same size wire with about ½ inch spacing between the two coils. Flood the turns with Duco cement to hold them from slipping. This coil form can be mounted vertically under the panel with two little angle "irons."

### Making the Receiver Plug

The plug is made from a discarded 5 prong tube. Knock out the bulb and remove the leads from inside the prongs. Then solder the aforementioned wires into the two heater and plate prongs and fill the shell with wax to support the wires.

The inside of the box can be lined with a tinfoil shield by shellacking the wood and foil and then, when they are tacky, spreading the foil smoothly in place all around the inside. Bring the foil at one end up and over the edge to make positive contact with the strip under the ground binding post. Paint the box any suitable color outside. Then fasten the panel and the "works" in place and you are ready to hook up to the receiver.

Remove the antenna from the receiver and attach to ant. post of the call-box. Then connect the ground post to the ground post of the receiver. Remove the detector tube from the receiver and place it in the call-box socket and insert the plug into the detector socket of the receiver. Turn on the receiver. This should light the call-box tube. If it doesn't, investigate the plug wiring from plug to socket.

When the receiver has warmed up you will hear the usual rushing sign. Turn the regeneration condenser about half way in and then slowly turn the tuning condenser. The dial should be a vernier for fine tuning. Don't be in a rush. Short wave work takes much more careful tuning than the broadcast band. If you hear a whistle in the speaker back off the regenerator until the signal clears up. It may be code or it may be voice.

### Tuning for a Police Station

With switch arm on A you will hear the calls lying in the 120 meter band and with it on B you can reach the bottom of the broadcast band with the police calls just below. In between will lie the code and phones of the amateurs working on the 160 meter band.

When a station is found the volume can be built up to its loudest point with the regenerator. Then carefully readjust the tuning condenser until you get the loudest and clearest signal.

Remember that with the call-box shown in the photos and herein specified the two variable condensers were of the 11 plate variety and the coils wound as specified. If you have no 11 plate condensers (cap. .00025) you may use 9 or 13 plate affairs but this will affect the windings of the grid coil somewhat. For a 9 plate tuning condenser you will probably have to have nearly 30 turns to the grid coil or for a 13 plate tuner the coil can be reduced to per-

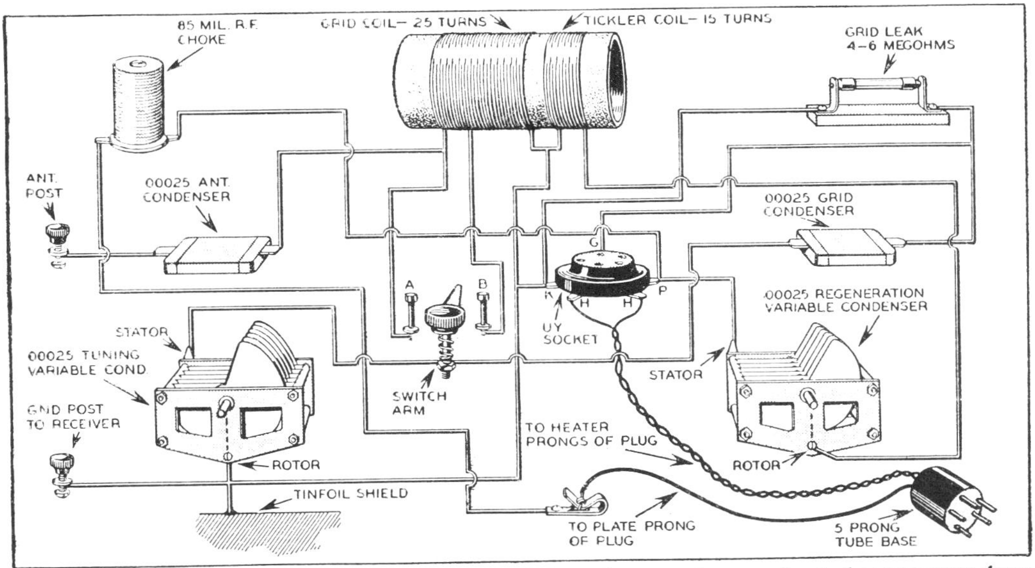


**Fig. 2. This picture diagram will make the hook-up easy for you. Run your wires according to the arrangement shown, keeping all leads as short as possible. Note that rotor of tuning condenser leads tinfoil on the inside of the cabinet.**

# *Police Call-box Can Be Assembled From Spare Parts in Radio Workshop*

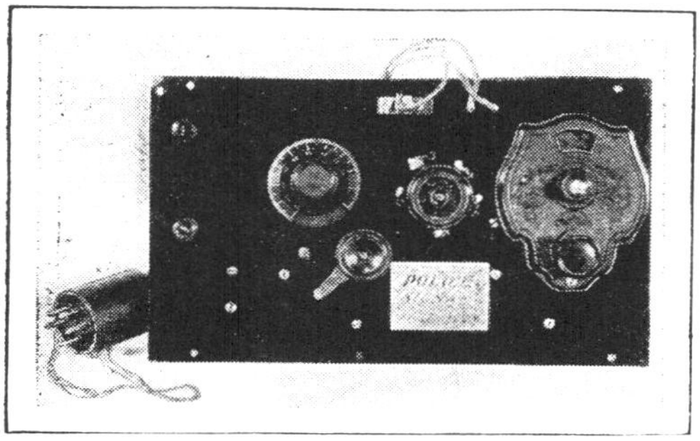

Top view of call-box. Tuning condenser dial is seen at right, regeneration condenser dial at left, tube socket in center.

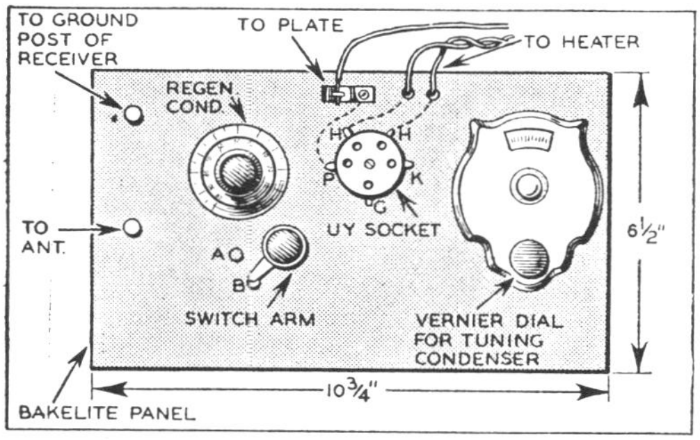


Fig. 3. This drawing, in conjunction with photo at top of page, will give you lucid idea as to efficient layout of parts.

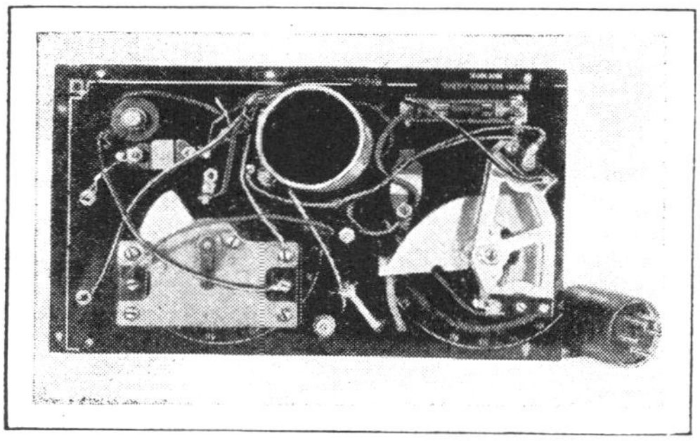

Rear view of the panel. The tuning condenser should have plenty of room to swing in. Coil form is seen at top center.

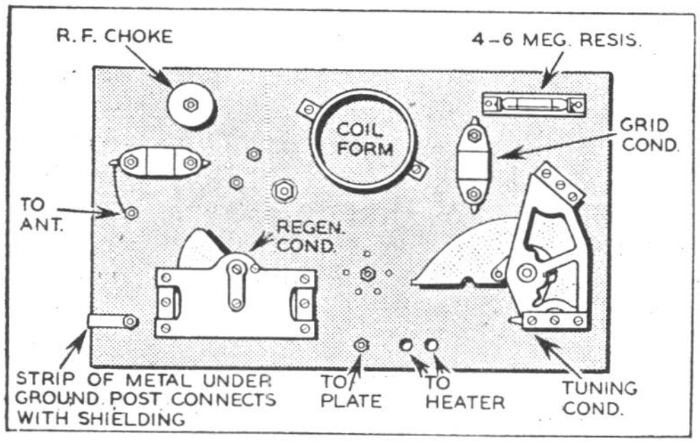


Fig. 4. Here's how you should arrange parts on the back of the panel. The strip of metal under the ground posts connects with the tinfoil in cabinet to provide ground.

haps 22 turns. This is only approximate and will call for some little experimentation. Better stick to the prescribed capacities, however, if possible, as the coil turns specified work admirably with the condensers specified.

Use a long antenna for picking up these police signals because with this arrangement no RF is used ahead of the detector and the usual short inside antenna or wall tape is next to useless. An antenna wire around 100 feet will do nicely.

### Stations You Can Listen For

Police stations using broadcast transmitters are banded together in wave length above and below the Amateur Band and switching over from A to B will enable you to bring in either section at will. Amateur telephony will also prove interesting when police calls lag. If situated away from cities the police calls will not be heard much except during evening hours but if you are situated in or near a city using this police system then you should be able to hear the calls at about any time.

## Wood Shaft Reduces Body Capacity

THE radio fan is often troubled with body capacity when the hand comes in contact with the tuning condenser dial.

Move the condenser back from the panel, and make an extension shaft with some insulating material. A piece of ¼" wood dowel rod will just fit the dial. Either make or buy a metal coupling connector having two set screws. Place this between condenser shaft and dowel, and tighten set screws. Six inches of such shafting will usually entirely eradicate any remaining body capacity.

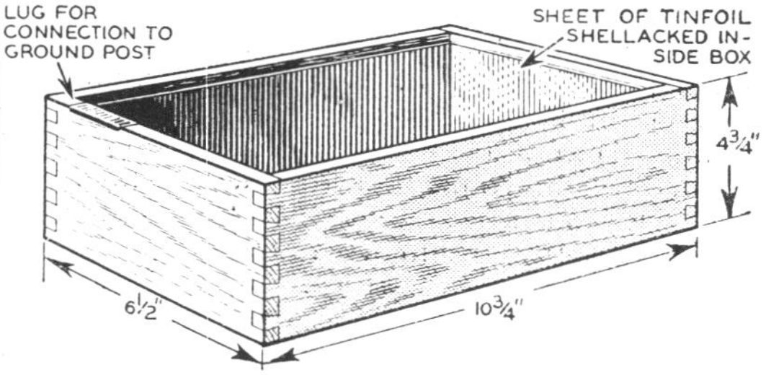

# Catch Coded Radio Messages With Paper Tape Recorder

SLOWLY turn the crank of this code recorder as the dots and dashes click in on your short wave receiver, and the coded messages will be neatly printed on paper tape. Turn the crank a bit faster, and you can record every click of the high speed machine transmitters, translating the messages at your leisure.

The action of the recorder is based on the fact that a small direct current will decompose potassium iodide, a salt obtainable at any drug store. If a current is passed between two wires touching a paper moistened with this salt, a brown line is drawn.

A tank for the potassium iodide solution is made from a cocoa can. Three wood dowel rollers are mounted to guide the tape through the bath, pins being used for bearings. Solder the pin heads on the outside, and coat the inside of the can with paraffin.

The reels are made from tin, with wood spools for the core.

Reel supports can be of heavy gauge sheet metal. The metal standard at the center supports the recording unit and paper tape guides. Two fine iron wires bent in the form of hooks are mounted on a hard rubber or wood block with machine screws. The two wires should not be over 1/32" apart where they touch the paper tape.

Ticker tape is ideal for the recorder. Rolls of thin wrapping paper can be cut into narrow rolls with a sharp hack saw.

A telephone receiver or even an old head phone will do for the necessary relay. Mount a spring armature over the receiver magnets, so that the contact points will close each time a code signal pulls down the armature. Connect the relay to the speaker terminals of the radio. Place a dry cell in series with recorder points and contacts, and the recorder is ready to use.

# Radio Power Pack Changes Increase Output Ninety Volts

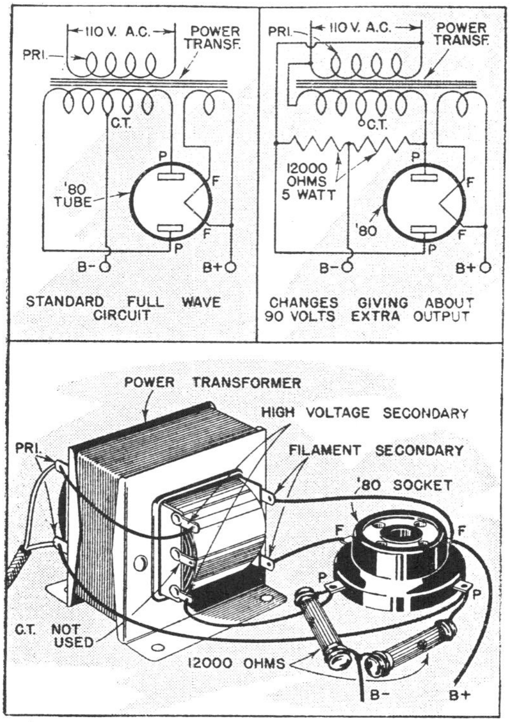


Auto-transformer connection shown above for power pack transformer gives 90-volt increase in output. Connections are made to center tap resistor as shown in lower drawing. If 10,000 ohm rheostats are used in place of the resistors, they can be adjusted to exactly same value with voltmeter.

HERE is an inexpensive method of increasing the voltage of your power pack by at least 90 volts. First, of course, determine your filter condenser ratings to see if they will stand the added voltage.

Old and new power pack circuits are shown at the right. The power pack transformer is connected as an auto-transformer, and primary and secondary voltages are thus added together. An artificial center tap, made up of two variable or fixed resistors, now replaces the secondary center tap connection.

Carbon pile variable resistances will permit a more accurate adjustment of the new center tap, since fixed resistors are seldom exactly the same value.

When using this power pack connection, a small capacity fixed condenser must be placed in series with the receiver ground, or no ground connection used at all. This is necessary because the negative side of the power pack output is now directly connected to the power line.

## Socket Antenna Fits Inside Set

HERE is a socket antenna system that can be included right inside the chassis of the receiver, with no visible outside wires.

Locate the 110 volt double cord line running from the switch to the power transformer. Untwist a part of this cord so you can get at both strands. Bare a spot in each strand about an inch long. Then to each bared spot solder one terminal of a .005 mf. fixed condenser. Any good receiver type fixed condenser will do.

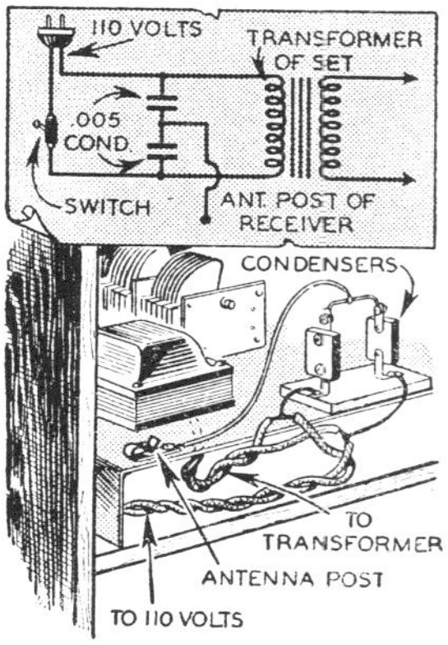


Connect the two remaining condenser terminals together with a short jumper wire. Solder a short piece of wire to this jumper and connect directly to the antenna post of the receiver.

## Connecting Tubes Without Sockets

HALF of an old hydrometer bulb nailed to a small wood baseboard will hold a radio tube tightly in an upside down position, when no sockets of the correct size are handy. Use tiny spring clips to make connections. Rubber balls cut in half will also do, if of the correct size.

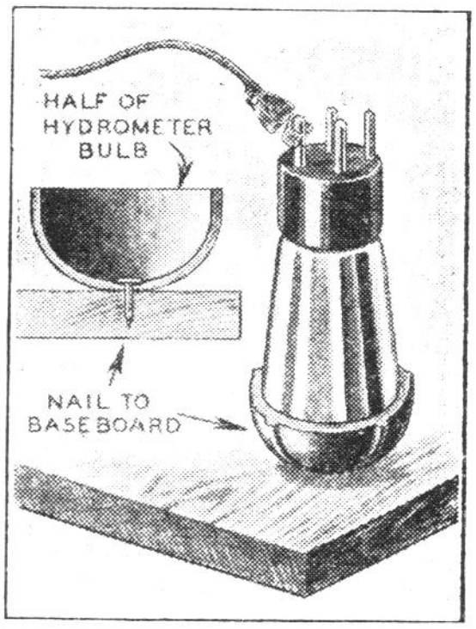


## Short Wave Vernier Spreads Bands

TWO broadcast size variable condensers, when connected in series, have a capacity equal to that of a midget short wave condenser. By using one as a tuning control, the other as vernier, the band-spreading feature is obtained. Keep the vernier condenser near a fixed setting.

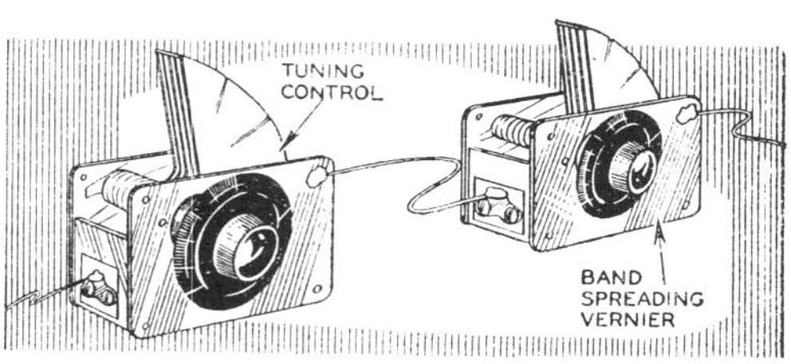

# Make Wet B Batteries for Radio From a Storage Battery

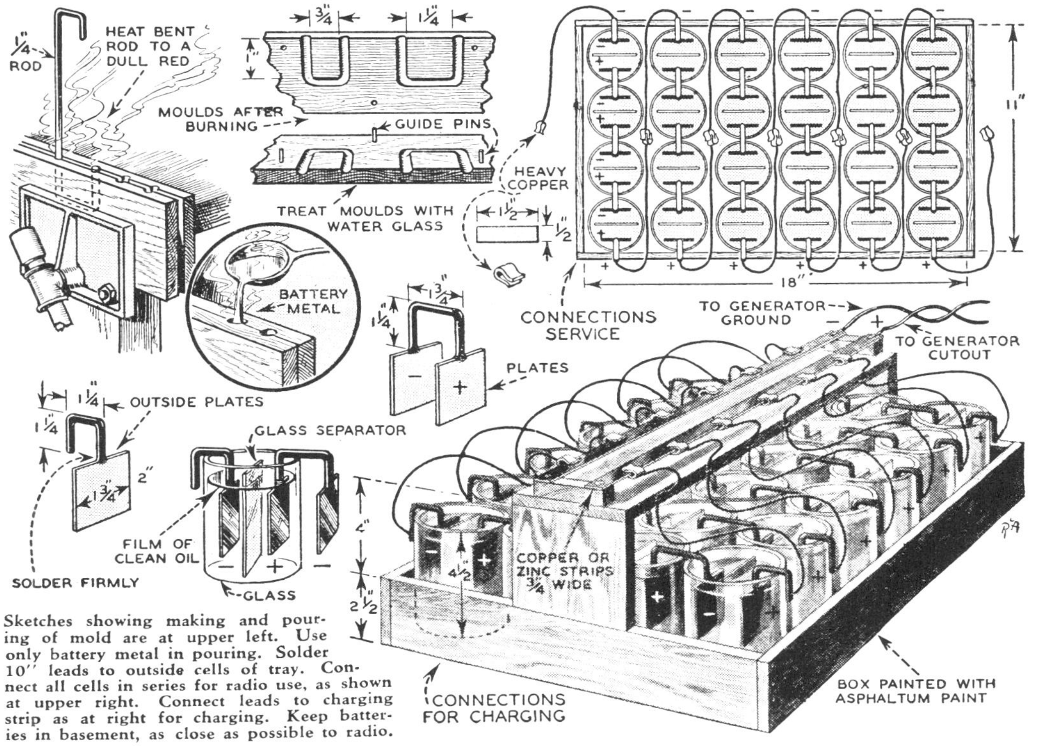


Sketches showing making and pouring of mold are at upper left. Use only battery metal in pouring. Solder 10″ leads to outside cells of tray. Connect all cells in series for radio use, as shown at upper right. Connect leads to charging strip as at right for charging. Keep batteries in basement, as close as possible to radio.

## Build a Heavy Duty Power Supply

THIS power pack will deliver approximately 500 volts at 200 milliamperes, enough to operate a small transmitter.

Use a power transformer rated at 600 volts each side of the center tap. Because of their cheapness, two 5Z3 tubes are used as half wave rectifiers. As such they will easily stand 1000 volts.

The filter condensers will have to stand up on 1000 volts. If electrolytics are used, connect two in series as shown.

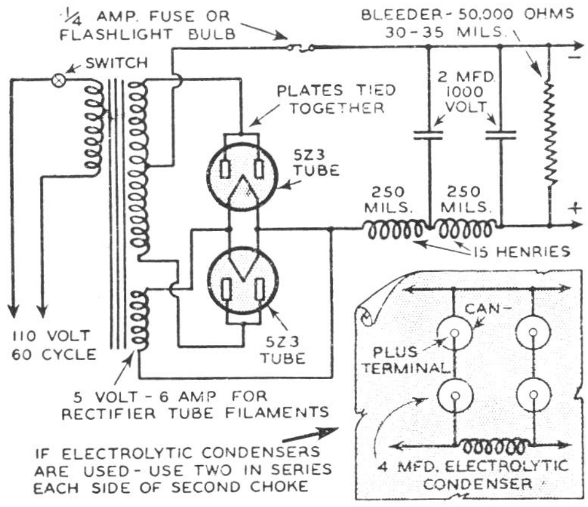


THIS wet type B battery entirely eliminates the B battery nuisance of old radio sets.

Saw off the plates of an old car battery—use good plates even if you have to buy a cheap, new 6-volt battery.

Allow the plates to dry thoroughly, then cut a section 1¾″x2″ from each corner.

Burn out the mold for the plate couplings, after setting in the aligning dowels. Place the mold in a vise and pour in melted battery metal. Use acid core solder in soldering the couplings to the plates.

Pint beverage bottles cut off 4½″ from the bottom are ideal for jars. Build a wooden tray for the jars. 24 units should test 60 volts when charged, each jar being 2½ volts normally.

Clean the glasses and plates in rainwater. Arrange the plates according to the diagram to secure the correct polarity. Separate the plates with glass panes.

Get sulphuric acid at your garage, and dilute with rainwater until it tests 1250 with a hydrometer. Don't pour water into the acid—it will explode.

Fill each cell until the plates are covered, then pour a thin film of clean oil on top.

For charging, the batteries are connected in parallel banks of four cells each, and connected to an auto generator.

# TESTED RADIO SHORT-CUTS

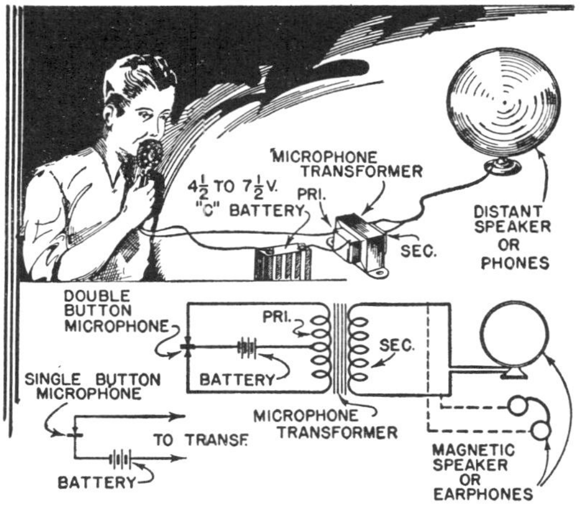


## *Amplifier Works Without Tubes*

PUBLIC address and inter-office communication systems giving loudspeaker volume without the use of any tubes can be set up by any radio experimenter.

The tube-less amplifier in its simplest form consists of a carbon microphone coupled through a microphone transformer to a magnetic speaker.

The transformer is a standard "mike to grid" type selected to match the microphone being used. A center-tapped primary is necessary if a double button microphone is to be used. The battery voltage is, of course, that recommended for the particular microphone at hand.

This amplifier will work with headphones as a "hard-of-hearing" aid.

## *Connect C Batteries in "Series Bucking"*

WHERE radio set specifications call for an odd value of grid voltage which cannot conveniently be built up by connecting small units in series, keep in mind this kink. Instead of connecting the batteries in series *aiding,* where the voltages of all cells add together, use a series *bucking* connection. Here one or more cells are connected in series with polarity reversed, these voltages being subtracted. In the example at the right, the use of this bucking connection allows two batteries to replace the three ordinarily required to give 18 volts.

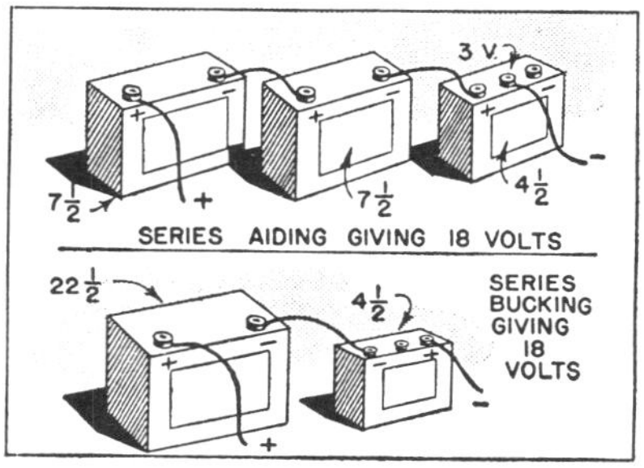


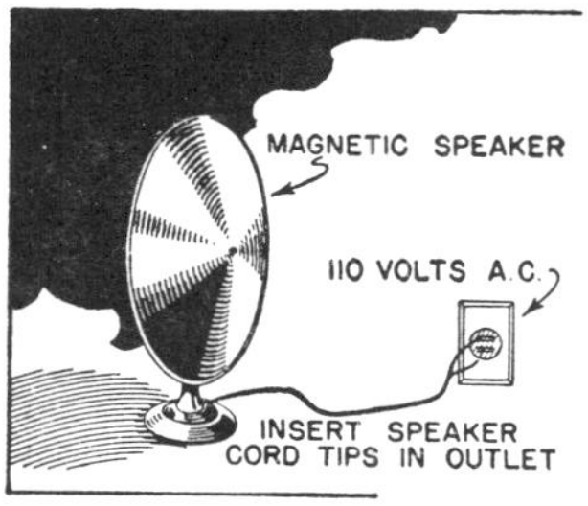


## *Check Magnetic Speakers With 110 Volts A. C.*

A MAGNETIC speaker can be checked for an open winding by plugging it into a 110 volt A.C. line. The speaker will produce a steady 60 cycle hum if the winding is good. When the speaker chatters, the armature is striking the pole piece, and should be adjusted. Two speakers can be compared for efficiency on low frequencies by comparing their hums on A.C. No damage will result, since the windings have a high impedance.

## *Add a Police Call Switch*

POLICE calls, aviation and some amateur stations can be brought in on many radio receivers merely by installing a single switch and resistor. The method is applicable to midget and standard size super-heterodyne sets having an intermediate frequency of about 456 kilocycles.

By using a double pole double throw switch connected as in the diagram at right, the set can instantly be changed over from broadcast to short wave. Dotted lines indicate the original set connections, and heavy lines the new connections. Short wave programs are received with the switch thrown to the right.

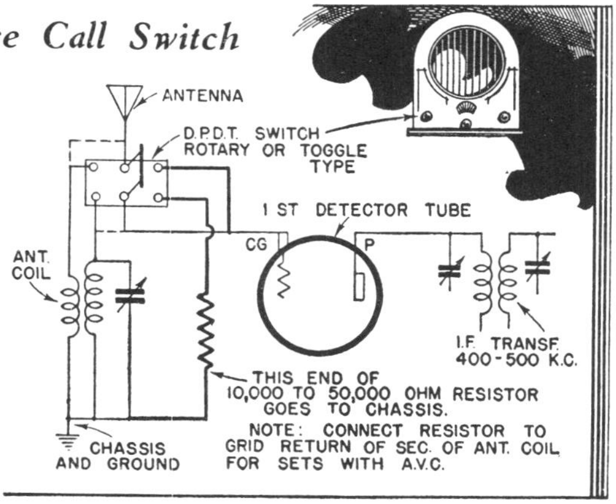

# Testing Set Locates Shorts, Open Filaments in All Tubes

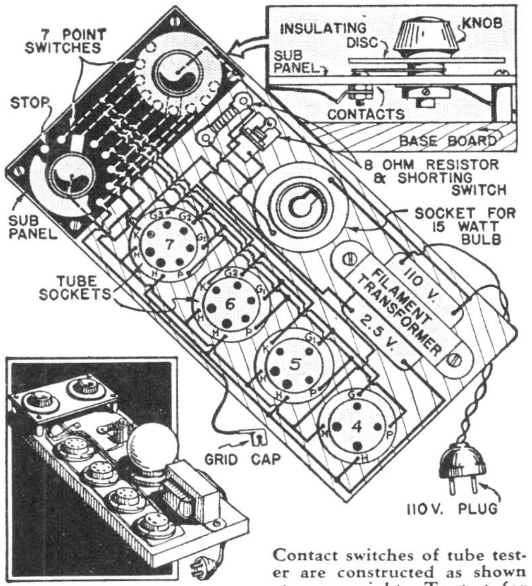


Contact switches of tube tester are constructed as shown at upper right. To test for shorts, spin dial of one as other dial is set on each contact in turn; bulb should light for only one setting in each case.

WHILE this radio tube testing device cannot give the exact operating characteristics of tubes being tested, it does tell whether it is safe to place them in the radio set, and whether the filaments are shorted or burned out.

The tester is shown in its most complicated form, made to accommodate all types of tubes. If only 4 and 5 prong tubes are to be tested, the layout can be simplified considerably.

The 2½-volt filament transformer will make the filaments of even 5 and 6.3 volt tubes glow. The shorting switch should be opened to place the 8 ohm resistor in series with the filament for the 2-volt tubes. It is closed for all other tubes.

The dial switches check all possible combinations of electrodes to locate any which are touching.

## Car Radio Draws Little Current

WITH this device connected to the car radio, when the car is in motion the charging rate of the generator is automatically raised 5 amperes when the set is turned on, and lowered as the load is cut off.

An old generator cut-out is connected from the set side of the radio switch to ground. When the radio is turned on, current flows through the coil, pulling down the armature and closing the contact points. A resistance placed in series with the grounded field lead is thereby shorted out and the generator charging rate increased.

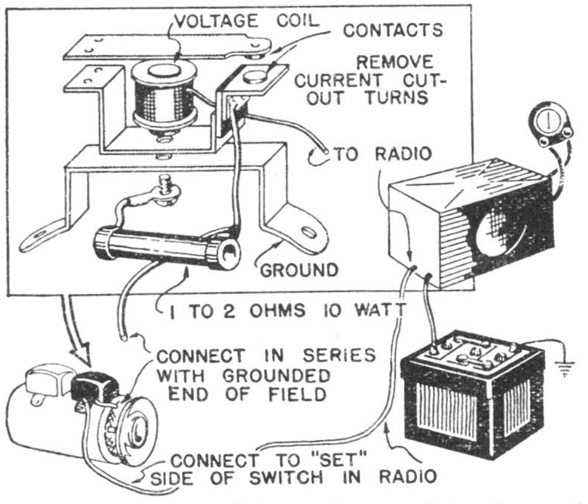


One end of cut-out coil is already grounded to frame. 2 ohm radio resistor connected from insulated contact to ground will drop charging rate 5 amp. on most generators.

## Connect Condenser Mike to Radio

WITH the highly sensitive new radio sets, a condenser microphone for public address work will often work satisfactorily in series with the antenna.

On electric sets this type of microphone can be hooked between the cathode and the grid of the detector tube. Attach to the 2nd detector tube on superhet receivers.

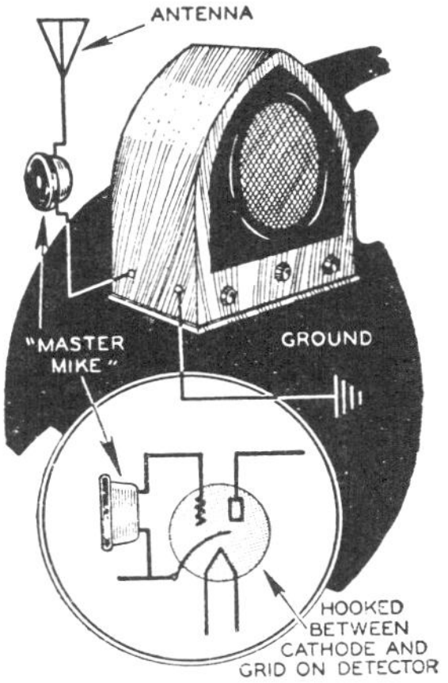


How "Master Mike" condenser microphone, described on page 51, is connected to the radio.

## Making Spring Type Binding Posts

ANY springy, stiff metal can be made up into spring type terminal posts for radio work. Cut out strips as shown at right, making the narrowed end small enough to fit through the slot. Punch out the slot with a cold chisel. A dozen clips can be made up in a short time.

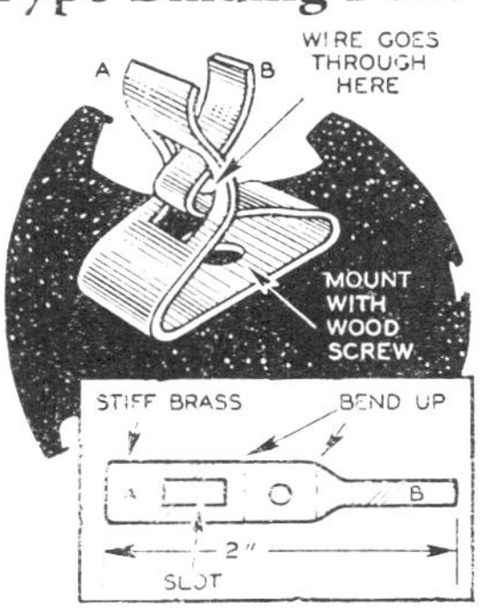


Terminal posts made by cutting out spring brass with tin snips and chisel, then bending.

# 1-TUBE Circuit for CAKE PAN Transmitter

Top and bottom views of one-tube cake pan transmitter. Filament switch is at end, tuning dial at top. Use telephone transmitter or any single button carbon microphone. Telephone coils were used here in place of audio transformer primary and microphone transformer; you may have to get a larger cake pan if standard radio parts are to be used.

THOUGH just a single tube circuit built into a tiny cake pan, batteries and all, this novel transmitter will broadcast through your own radio from any room in the house. Snap the single lead on to any convenient grounded pipe, flip up the filament switch, tune the transmitter to the setting of your receiver, and get "on the air."

Essentially the same circuit as that used for the small broadcasting station described on pages 64 and 65 of this book, this circuit uses the new type '19 tube in place of the two '30 tubes. The coil is exactly the same as before, but a smaller tuning condenser is used, to permit mounting into a compact unit.

Two small size flashlight cells connected in parallel supply 1½ volts, which works very nicely here for the filament power. Twenty of the pencil-type flashlight cells connected in series give 30 volts for the plate; many builders will prefer to use 45 volts, however, from a separate battery, to gain more volume with the midget set.

First line the inside of the cake pan selected for your chassis with some kind of insulating paper or cardboard. Solder threaded brackets into each corner, to which the heavy fibre or wood cover can be fastened.

In operating this set, the broadcast receiver is set at a point where no stations are heard, the volume is turned on full, and the transmitter dial is turned until a squeal is heard in the speaker. With the small tuning condenser used here, you will have to set your broadcast set on the lower part of the broadcast band between 1300 and 1500 kilocycles. Final adjustment for maximum volume can then be made with the tuning controls of either the transmitter or receiver. Once the two sets are tuned, you can operate the transmitter from any room.

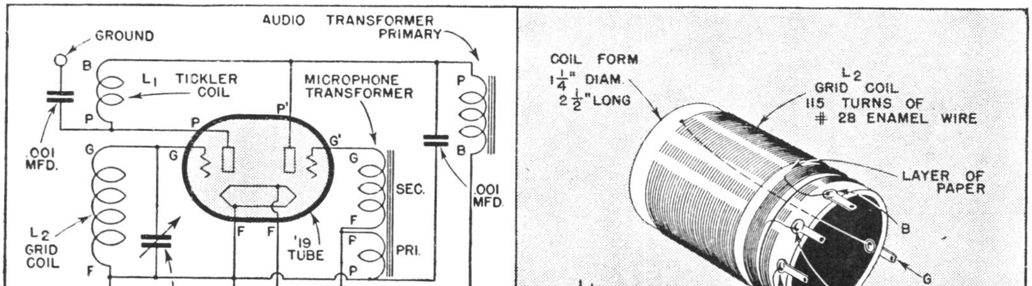


Circuit diagram and coil connections are given here, with lettered terminals to show correct coil polarity. Socket connections for '19 tube can be found in any radio tube chart or manual. Larger tuning condenser will extend band covered by transmitter. Wind grid coil first, place layer of paper over one end, then wind on tickler coil as shown.

# Install Running Board Antenna for Auto Radio

JUST what type auto antenna is best? It's a much discussed subject but, in the writer's opinion, the "capacity" plate type is the best for all around purposes.

The accompanying sketch shows a length of 4 to 5 feet, depending upon the length of running board or space available under the chassis. Cut a sheet of fairly heavy sheet zinc or copper the desired length and about 15 inches wide. Bend over each edge for a width of 1½ inches at right angles. These turned edges make the plate stiff longitudinally.

Now, with the bent edges down, fasten an oak block about a foot long and 3 in. high across each end of the plate by drilling the latter and using short screws from underneath. These blocks should first be boiled in paraffin at least two hours and hung up to drain and dry. This process makes them good insulators.

Now have two brackets made of flat iron about like those shown. Three holes should be drilled in the long portion and then fastened to the tops of the wood blocks with short screws. Also drill several small holes in the front end of the plate and weave in them the bared end of an insulated antenna lead-in wire. Solder this wire to the plate.

Bolt the bracket end to the under side of the running board or under the car away at least 3 inches from the metal portion of the car. That is the reason for the brackets. These of course must not be so long as to hang the antenna too low where it will be in danger of being swiped off by road obstructions in traveling. See that it is well insulated, both the plate and the lead-in wire, and if your radio is sensitive enough you will pick up programs with good volume.

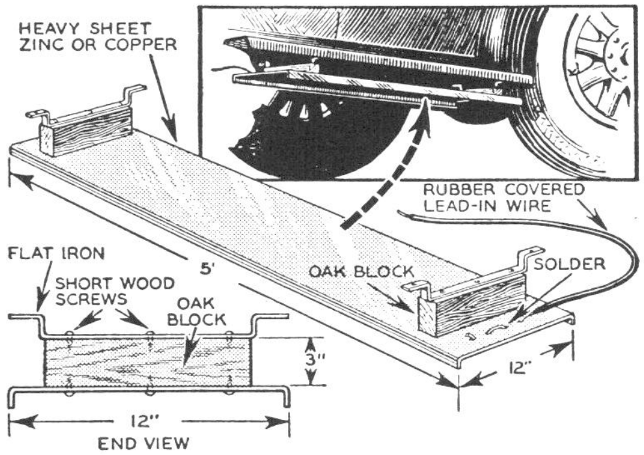


With the metal plate which acts as the antenna screwed to the block, bolt the two brackets to the under side of the running board or chassis, at least three inches from metal.

## "Dead Man" Made From 2x4's Holds Guy Wires of Radio Mast Securely

TO MAKE a "dead man" that will hold firm, no matter how much strain, proceed as follows. Cut out two pieces of 2 x 4 about two or three feet long, according to the size mast to be erected. Place them in the form of a cross and spike them solidly together. Then wind several turns of telephone wire about the middle and fashion a long loop from the ends.

Bury each "dead man" at least two feet deep, preferably more, and arrange it to lie at right angles to the direction of strain of the stay. Attach the latter directly to the loop; or better yet, use turnbuckles, and then the stay can be stretched taut. Pack rocks over the "dead man" and tamp the earth down well, and they will probably outlast the mast.

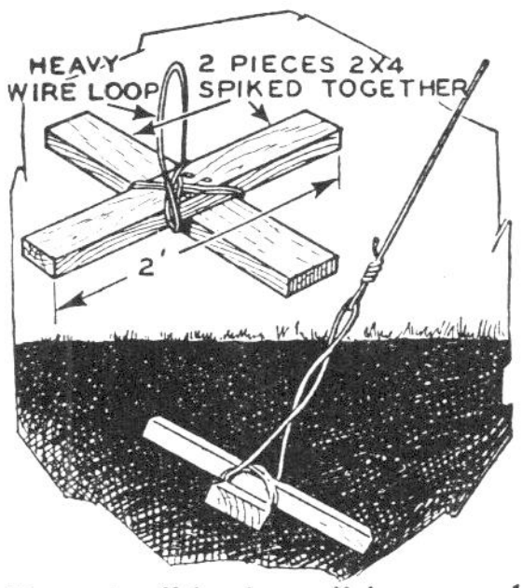


Bury the "dead man" in ground at angle shown in this drawing. Assembly of 2x4's also shown.

## Efficient Socket Made From Four Fahnestock Clips

PROBABLY every ham in the universe has at some time or other needed an additional socket and couldn't find one anywhere. Well, here's the answer to the problem.

Cut out a block of hard wood ¾ in. thick and 3 in. square. Spot the holes for the tube prongs by pressing the prong ends on a stamp pad and transferring these impressions to the center of the block. Then drill out these spots.

Now remove four Fahnestock clips from old B batteries. Drill a hole under the spring end for a screw hole and file a round slot in the end of each to match the sizes of the four holes.

Then solder a narrow strip of very thin spring metal from the screw hole down to the slot and bend down so the strip hangs down in the holes at right angles to the clip, as shown in the drawing.

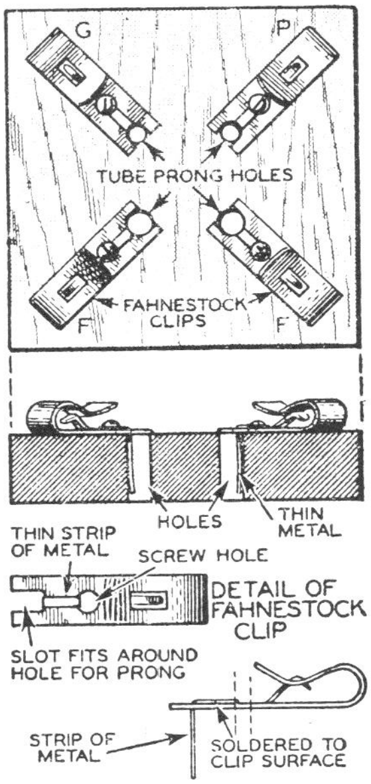

# Making Useful Radio Parts From Phonograph Records

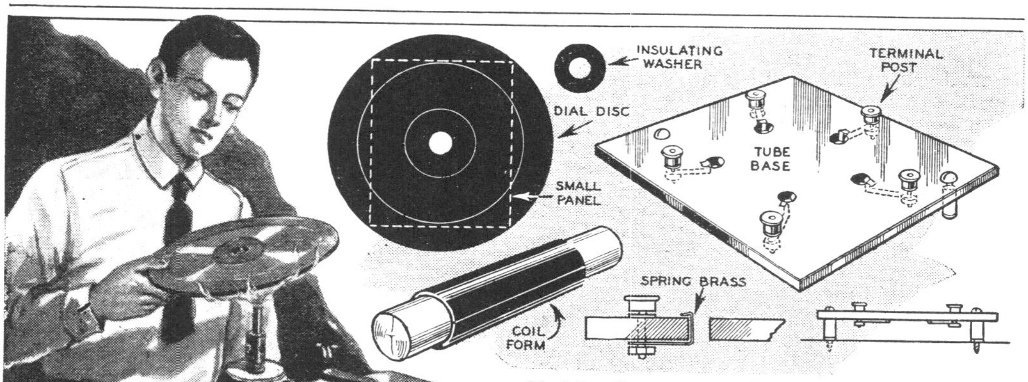


Useful radio parts are easily made out of old phonograph records. Bunsen burner, oven, or any heat source can be used to soften records. Do not use too much heat.

OLD disc phonograph records can be turned to good account as tube bases, coil forms, insulating washers, small panels, and dials.

Heat the records gently in an oven, placing them on a piece of flat tin or sheet iron. When they are soft, but not melted, they are cut to the desired shape. Records may also be heated over a bunsen burner.

Bend pieces of the records over desired diameter cylinders to make coil forms.

Cut out insulating washers with a panel cutter or razor blade attached to a compass.

Tube sockets should be cut 2½" square. Spot the tube holes and drill slightly larger than the tube prongs. Drill holes for binding posts at the corners, and run thin strips of spring brass from each binding post up through the tube prong holes.

## Simple Remote Control for Volume

TO REDUCE in volume or cut out entirely an annoying radio program from any remote point in the house, cut the aerial wire, and run twisted leads to the desired location of the remote control. Connect leads to a 23 plate (.0005 mfd.) variable condenser mounted in a neat box. This added capacity will not affect quality of signals.

## Hang Watch on Mike to Modulate

FREQUENTLY, when calling a fellow amateur, the "ham" has occasion to move from the transmitter, leaving the carrier wave unmodulated. A watch hung on the microphone will produce a steady signal.

## Build '27 Doubler "B" Eliminator

HERE is a full wave 110 volt a.c. "B" battery eliminator, using the voltage doubler principle, which will deliver around 135 m.a. at 200 volts plate supply, without any power transformer. Two easily obtained type —27 tubes here give the same results as the new 25Z5 tube.

Condensers C are common paper condensers, which need not be rated at more than 200 volts for an input voltage of 110 volts. These condensers double the line voltage.

Increasing the capacity of the condensers C increases the output.

Since one side of a power line is always grounded, this eliminator will not work when the radio is connected directly to ground. Place an .00025 mfd. fixed condenser in series with the radio set ground.

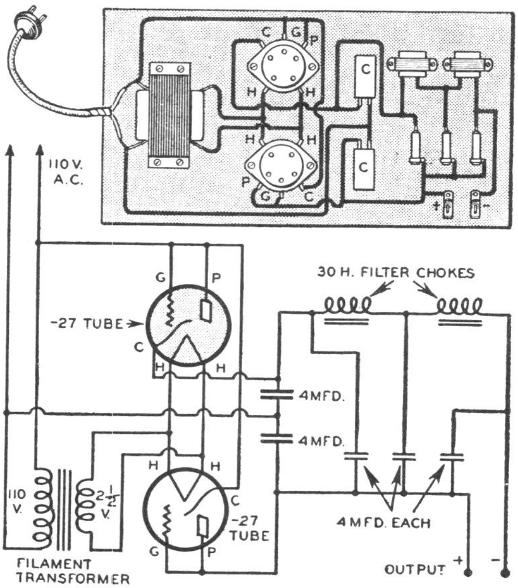

# Fascinating Stunts Teach Electricity

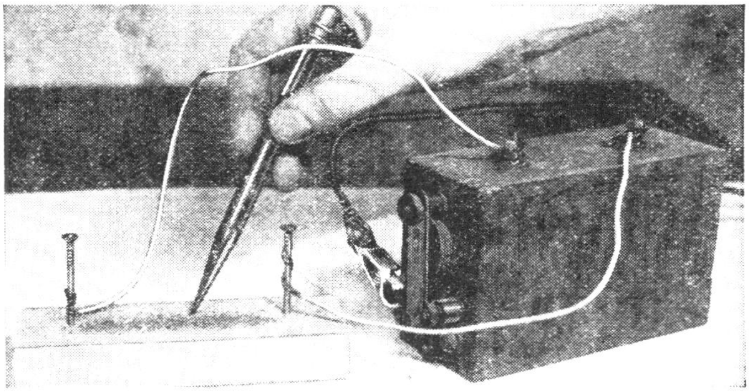

You'll see a fascinating display of electrical fireworks when you hook up an old Ford spark coil to a couple of nails. Iron fillings, sprinkled between nails, sparkle brilliantly.

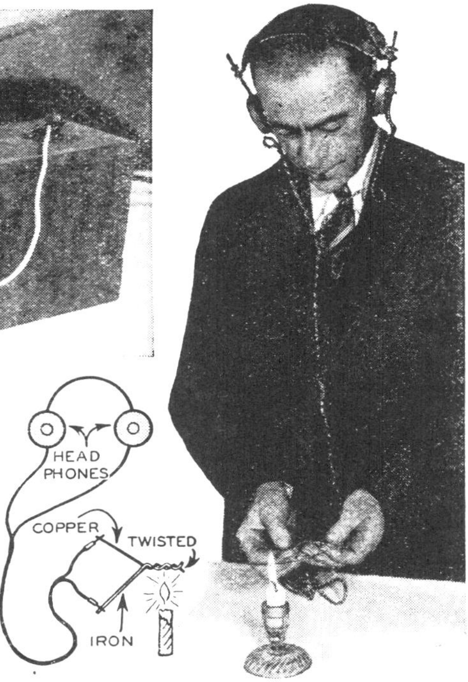


Principle of the thermocouple is illustrated when you twist a copper and an iron wire together and heat the joint over a flame. Tips of headphone cords touched to wires show current is flowing by loud click. Heat generates the "juice."

The function of a fuse may be studied by hooking strip of tinfoil across 110-volt line as shown. Turn on light—pfssst goes the fuse.

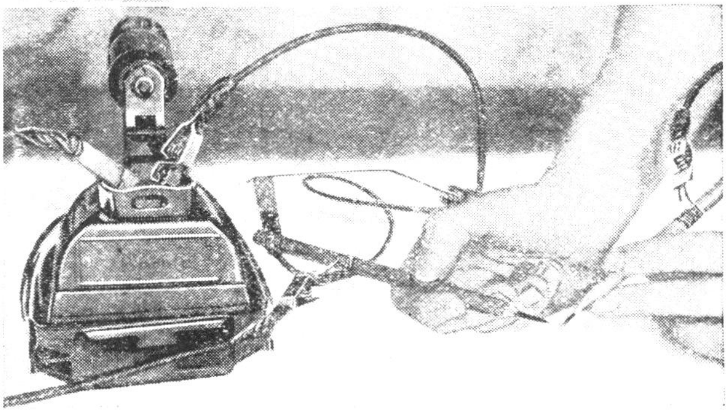

More fireworks. Aluminum powder, which comes with aluminum paint, is blown into a flame with a glass tube or straw as illustrated here. The small particles burn like myriads of shooting stars. Try this stunt for some startling entertainment.

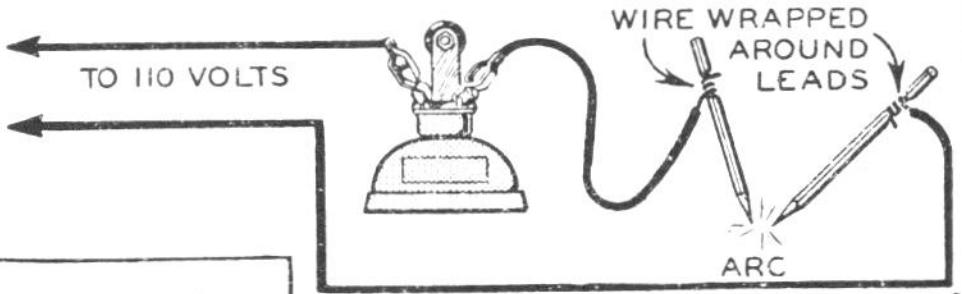


Two lead pencils and iron hooked up as illustrated show how an arc light works. First touch leads together, then separate slightly to start the arc.

# Efficient Sending Key Made From Door Bell Pushbutton

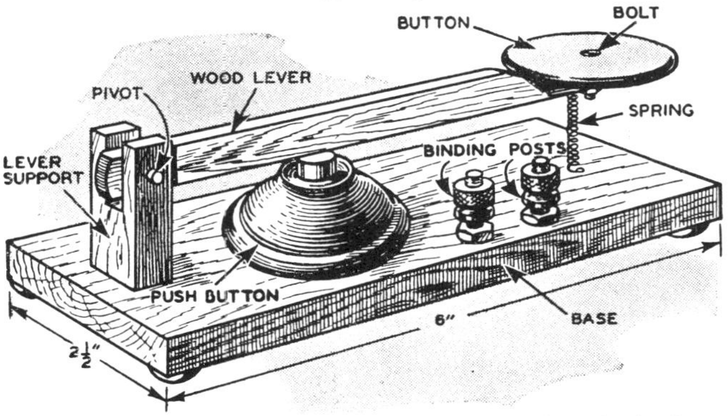

Pushbutton and key lever are mounted on baseboard as shown in drawing.

AN ORDINARY pushbutton, mounted on a baseboard and keyed as shown in the accompanying drawing, will make about as good a transmitting key as you could possibly want. At one end of the baseboard mount an upright, with a fork cut in one end. Then cut out a wood lever about ⅝ in. square and six inches long and insert the end in the upright fork, hinging it with a pinion. On the other end of the lever bolt a button to serve as a knob.

Next place a pushbutton on the base so the knob comes directly under the lever when the latter bears down on it. A spiral spring holds the lever in normal position as illustrated. Leads from the pushbutton are wired to the binding posts. Pressing down on the lever closes the pushbutton contacts to make the dots and dashes. Connect one pushbutton terminal to each binding post.

## Testing Transmitters With This Dummy Antenna Prevents Interference

THE transmitting amateur can make good use of a dummy antenna for testing purposes. It really should be used at all times when not actually on the air as it will greatly reduce interference to others and give just as satisfactory results for trying out different adjustments of the transmitter or new outfits and parts.

The accompanying photograph and diagram show how to assemble a simple apparatus to accomplish this. Use any suitable baseboard 8 to 10 inches square. At one side mount the antenna coupling coil made of 8 to 12 turns fairly heavy wire. Then at right angles mount a second coil of any reasonable number of turns of smaller wire tapped every few turns as indicated. The rheostat should be of from 25 to 30 ohms resistance. Then at the rear of the baseboard mount a .00025 or .0005 variable condenser and an antenna radiation meter, or, failing the latter, a flashlight bulb or an old 199 type tube and socket. Connect the parts as illustrated in the diagram.

To use this dummy proceed as follows: Provide a support for the outfit so the coupling coil will be a few inches from the primary coil of the transmitter exactly as any antenna coupling coil would be placed. Start the transmitter and close the key circuit. Now by means of the tapped coil, varying the rheostat resistance and the capacity of the condenser, obtain a reading in the antenna indicator that will approximate that of the regular antenna when in use. Mark the positions of the various parts and that's all there is to it.

Whenever the transmitter is to be tested simply remove the regular coupling coil, substitute the dummy antenna instead and you can make all your trials and adjustments without causing overdue crowding of the air and interference with transmissions by other amateurs.

It is good technical practice as well as amateur operator's courtesy to use a non-radiating antenna for transmitter experiments which do not require a radiating antenna system. There is then no interference with other stations while the transmitter is being given its preliminary adjustments.

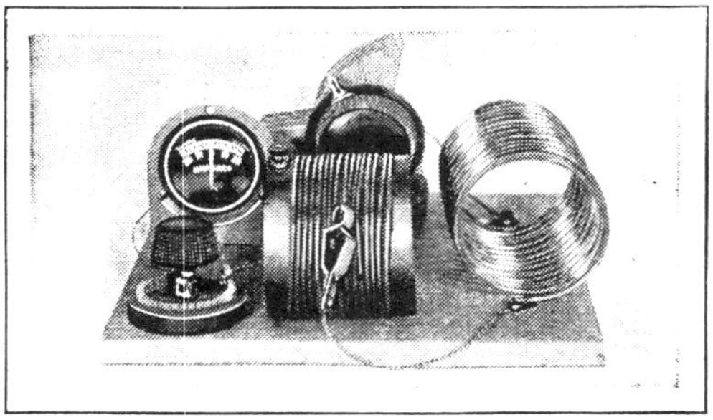

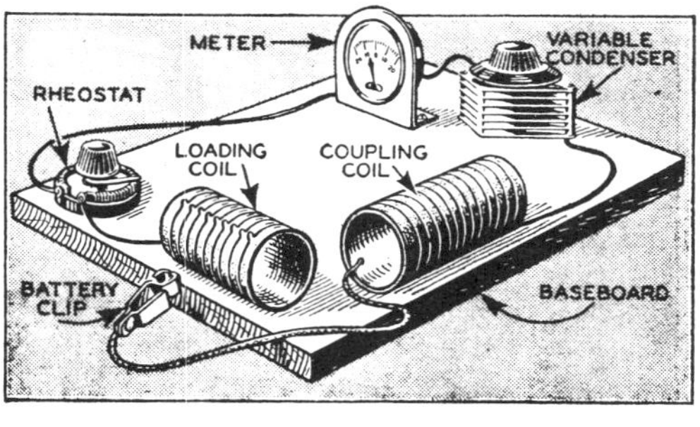

Photo above shows completed "dummy" antenna. The drawing illustrates arrangement and wiring of the parts.

# How to Obtain More Service From Discarded Dry Cells

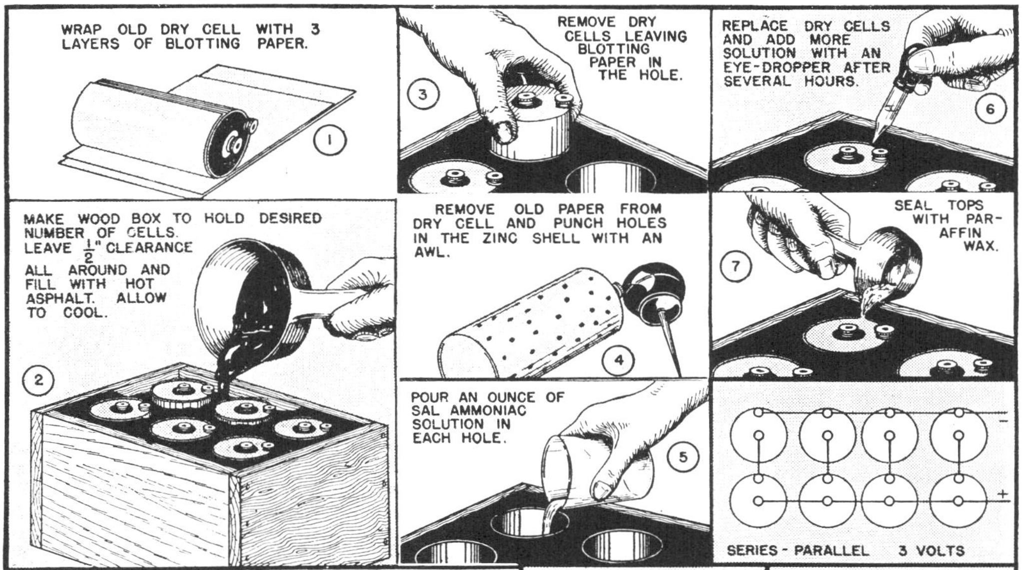


Above sketches indicate procedure to be followed in bringing discarded dry cells back to life. Four ounces of sal ammoniac can be obtained at drug stores for about 15c.

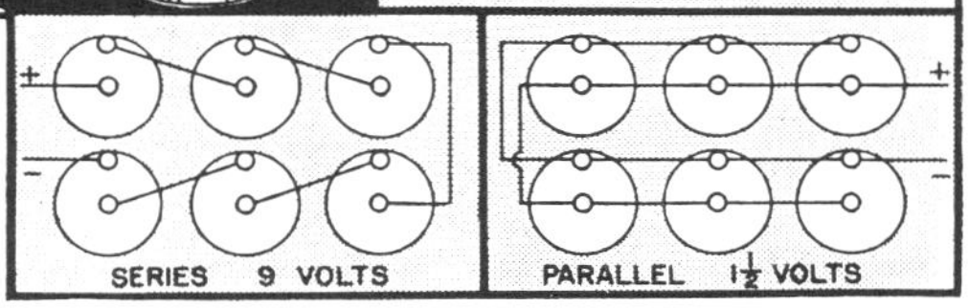


Series-parallel connections are used for tube filament power; series for B or C batteries; parallel for high current.

EVEN though a dry cell is completely dead, the rejuvenating procedure covered in the above sketches will make it deliver rated voltage just as long as there is any zinc left in the case. Four of the cells hooked in series make a good hot shot battery for ignition purposes.

The size of box to be used will of course depend upon the voltage desired, each cell being rated at 1½ volts. The asphalt used between cells can be obtained in chunks from any paving contractor, and melted in an old kettle. A thin layer of asphalt should be poured over the bottom of the box before setting in the cells.

Add four ounces of sal ammoniac to a quart of water to obtain a saturated solution, and pour a few ounces into each cell pocket. Several hours after replacing the cells add more solution to bring the liquid level nearly to the tops of the cells.

## Tubes Replace Filament Rheostats

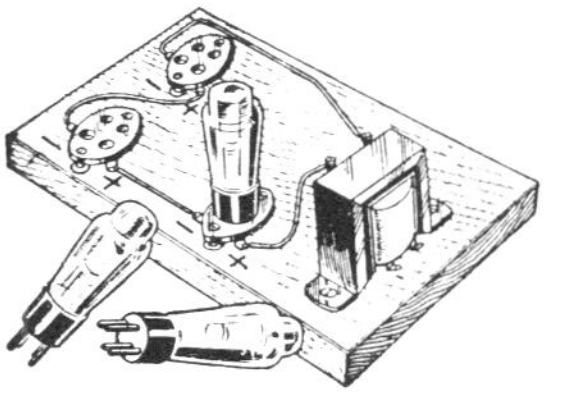

CONNECT filaments of spare tubes in series to replace resistors in dropping battery or transformer voltages for a tube filament. Tube filament resistances may be found in any radio tube chart or manual.

## Clothespin Makes Handy Switch

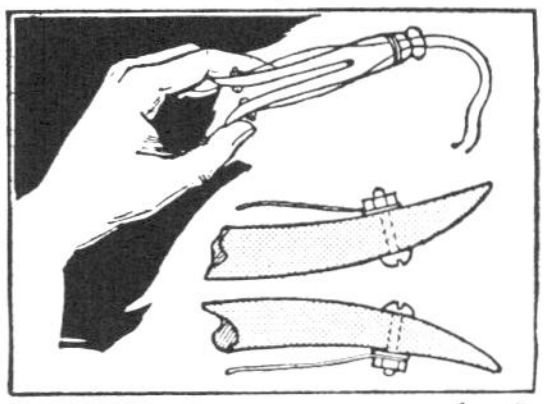

MOUNT two brass bolts with the heads facing each other at the ends of the tines of a clothespin. Run lead wires to the bolts, and you have a handy temporary switch which can be used for flashing signal lights or operating a buzzer circuit.

## Operating Radio Without Aerial

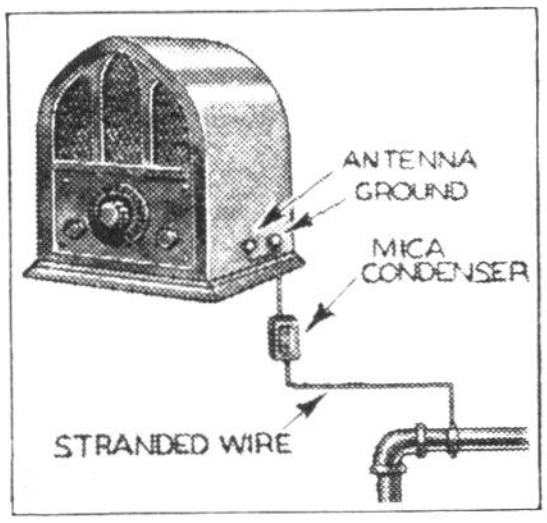


Connect radio as shown to bring in programs without aerial. Use any size condenser in ground lead as shown above.

EXCELLENT reception is obtained on many radio sets without using an aerial, if the antenna and ground binding posts of the set are connected together with a short wire. A mica condenser is placed in series with the ground lead.

# Smallest ALL-ELECTRIC RADIO HAS ONLY ONE TUBE

Your all-electric radio is ready for use when plugged in and table lamp is turned on. Tune with top dial, control regeneration with lower dial.

*by* THOMAS A. BLANCHARD

HERE'S the world's smallest all-electric radio receiver—a one-tube set that is not a bit harder to build than an ordinary one-tube battery radio. Best of all, this set will cover both the short wave and broadcast bands, bringing in police calls, phone broadcasts, code and phone amateurs, and perhaps even a foreign station or two under good atmospheric conditions.

The secret of the success of this tiny set lies in the tube used—a dual purpose type labeled 12A7. One-half of the tube is equivalent to a '38 pentode used as detector, and the other half a rectifier.

Instead of using an expensive special cord with built-in resistance to cut down line voltage for the tube filament, a novel current-saving arrangement has been devised. A table lamp with 40-watt bulb is plugged into a receptacle mounted on the set, this dropping the voltage to just about the right value for the filament. To shut off the radio, simply turn out the table light.

While too much should not be expected

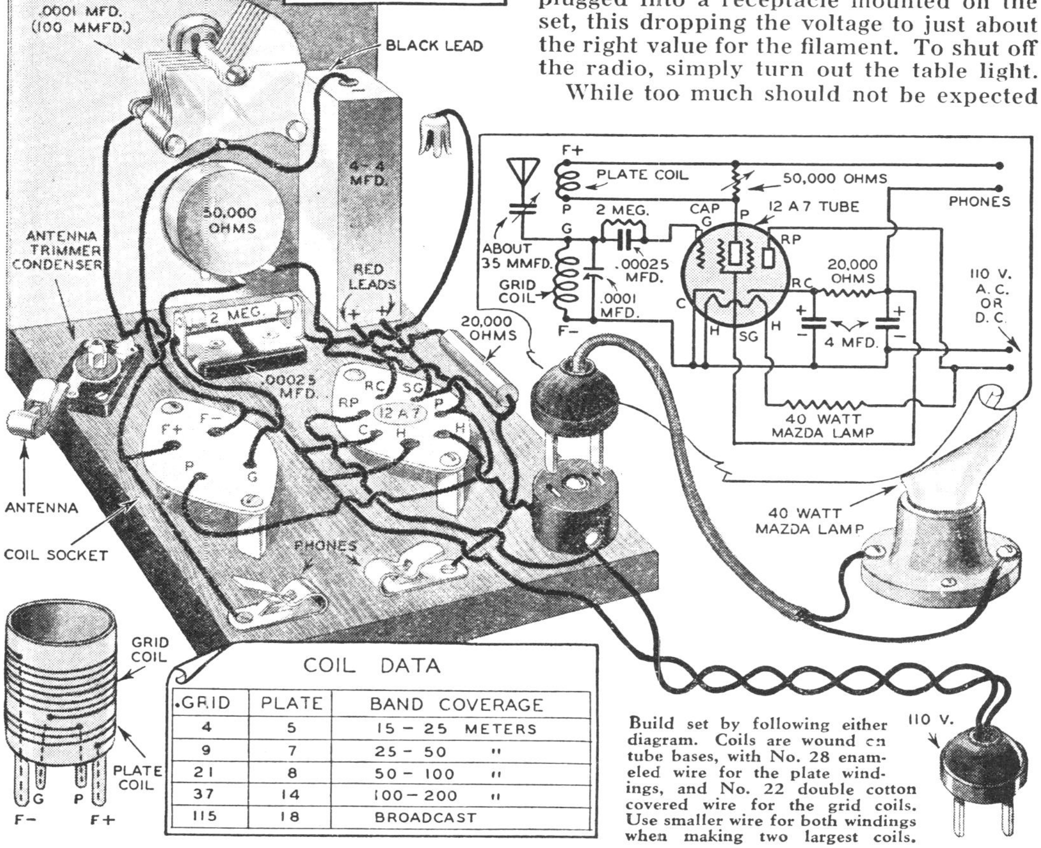


COIL DATA

| GRID | PLATE | BAND COVERAGE |
|---|---|---|
| 4 | 5 | 15 – 25 METERS |
| 9 | 7 | 25 – 50 " |
| 21 | 8 | 50 – 100 " |
| 37 | 14 | 100 – 200 " |
| 115 | 18 | BROADCAST |

Build set by following either diagram. Coils are wound on tube bases, with No. 28 enameled wire for the plate windings, and No. 22 double cotton covered wire for the grid coils. Use smaller wire for both windings when making two largest coils.

# 1-Tube Amplifier Brings Speaker Volume to Headphone

Photo of one stage amplifier using type —47 tube shows extreme simplicity of unit. Any good grade audio transformer may be used for input. Connect 5 prong socket as shown. Filament source may be either 2½-volt transformer or 6-volt storage battery with suitable rheostat. With battery, center-tapped resistor is not necessary; 450 ohm resistor is then connected to negative side of filament line.

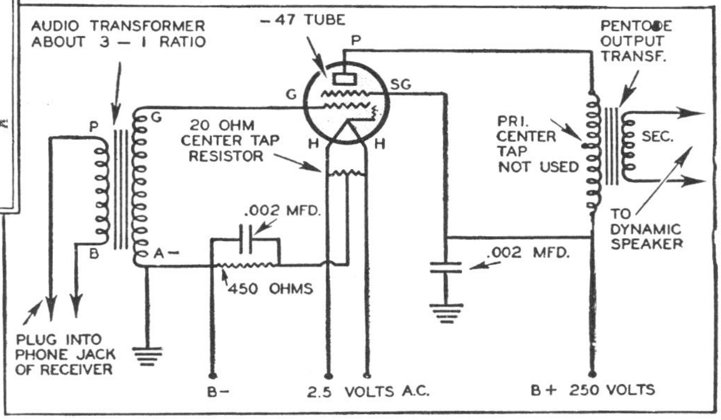


SOONER or later short wave fans having headphone receivers yearn for loudspeaker volume. It is much better to build a separate amplifier such as this which may be plugged into the phone jack of the set when needed. Speaker volume will not be terrific, but loud enough for a small room.

The amplifier should operate from a separate power supply, either batteries or a power pack. It uses a single type —47 pentode tube, which requires only parts generally to be had in the radio scrap box.

## Smallest All-Electric Radio Receiver Has Only One Tube

of this set in the short wave bands, it will give excellent headphone reception on the broadcast bands. With careful adjustment of the antenna trimmer condenser each time a plug-in coil is changed, a surprisingly large number of short wave stations can be picked up. On strong local stations a magnetic speaker may be used with surprisingly good volume.

For ordinary reception the trimmer condenser need be set only once, at an average value which will give best results for all of the coils.

On the broadcast band this set picked up local stations with no antenna at all. Powerful distant stations were received on a 6-foot long antenna connected directly to the stator of the tuning condenser, while a 100-foot antenna was used for reception of short wave and weak broadcast band signals. This long antenna was connected through the antenna condenser, this being either the screw-adjusted type or a homemade bookplate condenser like that described on page 37 of this book.

The resistance required to drop line voltage from 110 volts to the required value of 12 volts for this tube is about 360 ohms. A 40-watt bulb will do very nicely—any larger size bulb will blow out your tube.

All parts are mounted on a 3½"x5" metal panel and on a wood base of the same size. Care must be taken that the panel does not come in contact with any grounded objects such as steam or water pipes. Though the tube will not be damaged, one winding of your coil will burn out. Reversing the plug in the light socket will give the proper polarity to ground the set through the electric light line. Do not touch the metal panel and a grounded object at the same time unless you are sure polarity is correct.

Follow either the circuit diagram or the pictorial sketch when assembling the set, depending on which is easier for you. Tube socket connections are shown in the pictorial sketch, and should be followed exactly. Basemount sockets will be easier to connect, but wafer sockets are cheaper, and may be raised from the baseboard as shown.

A low hum may perhaps be heard in the phones with the 4-4 mfd. filter condenser. If this becomes too annoying, it can be reduced by using an 8-8 mfd. condenser instead.

The parts which you will need for this set are: 12A7 tube; 7 prong socket; 4 prong socket; 100 mmfd. midget variable condenser; 35 mmfd. antenna trimmer condenser; tuning dial; 50,000 ohm potentiometer; knob; .00025 mfd. fixed condenser; 2 megohm carbon resistor (1-watt); 20,000 ohm carbon resistor (1-watt); 4-4 or 8-8 mfd. electrolytic filter condenser; grid cap; 3 Fahenstock clips; hookup wire; single-screw mount receptacle; extension cord with plug; 3½"x5" panel; baseboard; four tube-coil forms; small spool of No. 22 D.C.C. wire; small spool of No. 28 En. wire; headphones.

# Midgy Jr.—Complete Vacation

WELL, radio fans, here's the latest in receivers. Here is a job that weighs less than four pounds including batteries. Lug it around like a camera or carry it in your brief-case, overnight bag, car pocket or even in a large overcoat pocket. It's all the same to Midgy Jr.

For local reception the only accessory is a pair of headphones. For DX work you will, of course, need a coil of flexible wire for an antenna and a second short piece for a ground connection. So there you are.

First clean off all tax tags, etc. from a cigar box and sandpaper the wood where necessary. Stand the box on end so the hinges are at the right. Across the box, 2 7/8 inches down from the inside of the top, glue in a horizontal partition as shown for the A and B batteries. The lower section is for the receiver parts in the circuit, which are arranged as shown in the photo and drawing.

The parts can be wired with small flexible wire such as No. 22 or No. 24 covered with spaghetti to insure against shorts. Solder all joints where necessary and make sure of all bolted connections.

Winding the coils is a simple matter provided the accompanying winding diagram is carefully followed. Use No. 24 DCC wire, winding on tube bases and using the large and small prongs exactly as shown. The large prongs are for the tickler coil connections and the small prongs for the grid coil.

Solder the ends to windings into the prongs and flood the turns with Duco cement. The number of turns shown will suit the various bands although it may be necessary to add or subtract one or two turns from the grid coils to suit just the points in the bands you may wish to reach. This is a matter of cut and try.

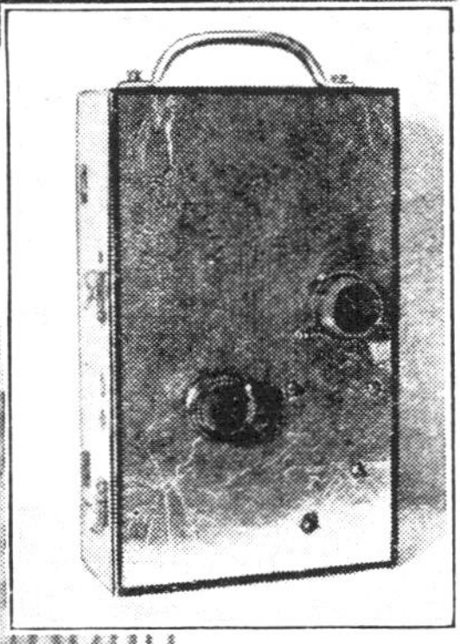

Completed receiver, shown above, may be carried like an ordinary camera. When set up in camp, in boat, or in auto, a small antenna should be strung up to bring in clear, loud signals on headphones.

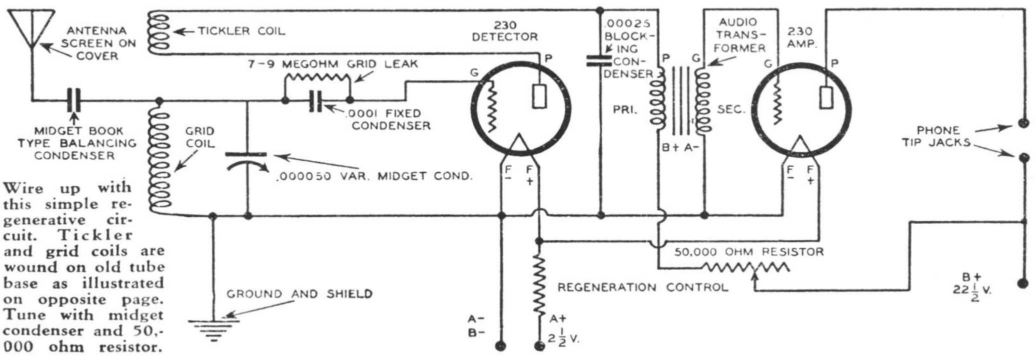


Wire up with this simple regenerative circuit. Tickler and grid coils are wound on old tube base as illustrated on opposite page. Tune with midget condenser and 50,000 ohm resistor.

# Receiver in a CIGAR BOX

On the inside of the cover fit in a piece of copper window screen wire for an antenna for local use. Connect this with a short piece of flexible wire to the antenna post on the outside of the box through a small hole for the same.

Then shellac the outside of the back, where the tuning condenser and variable resistor shafts come through and press on a sheet of heavy tinfoil to cover the entire surface. Plenty of shellac will make the foil adhere nicely.

Cut around every bolt and screwhead and projecting part so the foil will touch no metal part of the circuit. Then connect one corner of the foil to the ground post. This shielding is necessary to kill hand capacity that would otherwise make tuning very difficult.

Put the two A battery cells (flashlight) in the case at the right and the 22½-volt B battery at the left. A suitable hold down can be made by jamming a slender stick of wood across the whole battery assembly between the sides. Then connect the A and B wires to the batteries, put in the phones and you are ready to go.

As before stated, if there are no local stations, antenna and ground wires will be needed. For real portability, it is possible

*(Continued on page 96)*

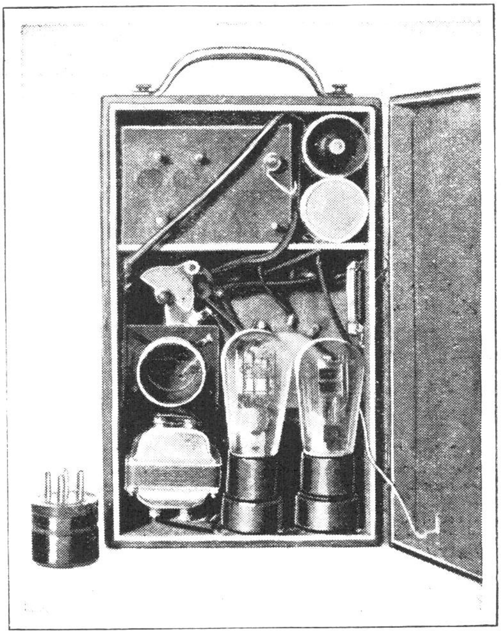

**Arrange parts in cigar box as illustrated in this photo.**

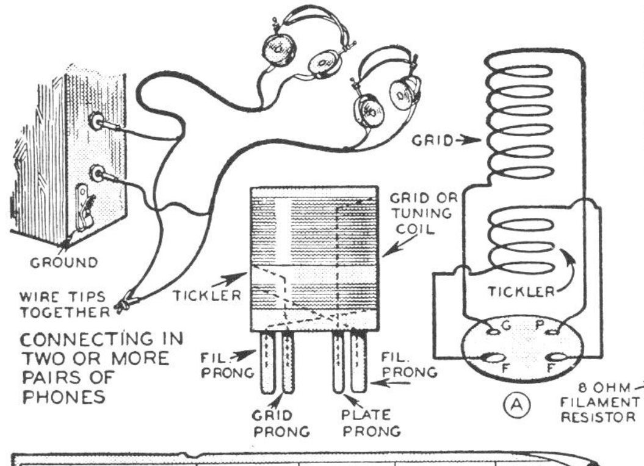


| | BAND METERS | WIRE NO 24 D.C.C. | GRID TURNS | TICKLER TURNS |
|---|---|---|---|---|
| THESE WAVE LENGTHS | | | | |
| ARE ONLY APPROXIMATE | 300-500 | " | 145 | 55 |
| AS CONDITION OF PARTS | 200-300 | " | 90 | 35 |
| AND ASSEMBLY WILL | 125-190 | " | 42 | 18 |
| GOVERN USE OF COILS— | 65-85 | " | 20 | 10 |
| TRY ONE AND SEE WHAT | 35-45 | " | 9 | 7 |
| IT WILL DO-THEN ANY | 16-25 | " | 5 | 6 |
| DIFFERENCES CAN BE ADJUSTED IN WINDING OTHERS | | | | |

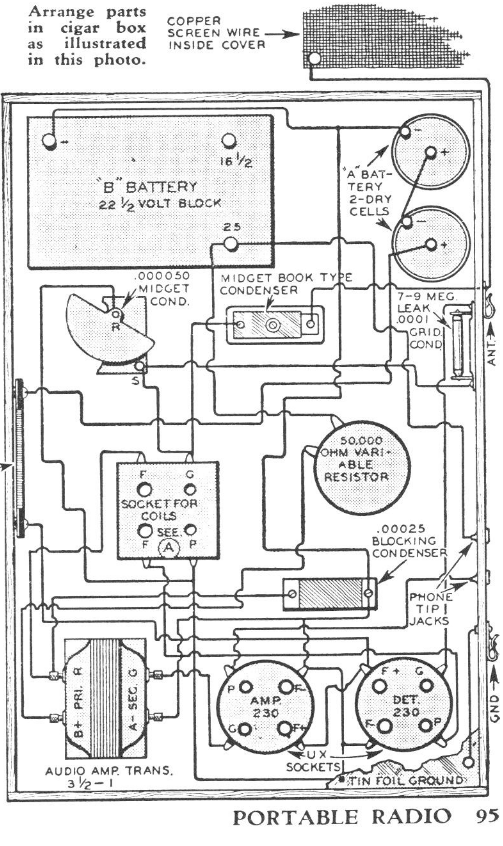


**Diagram at right shows how parts are wired up in the cigar box. Copper screen antenna for local reception is secured to inside of lid, and connected to the antenna post.**

# A Quick-Action Coil Changer for Short Wave Receivers

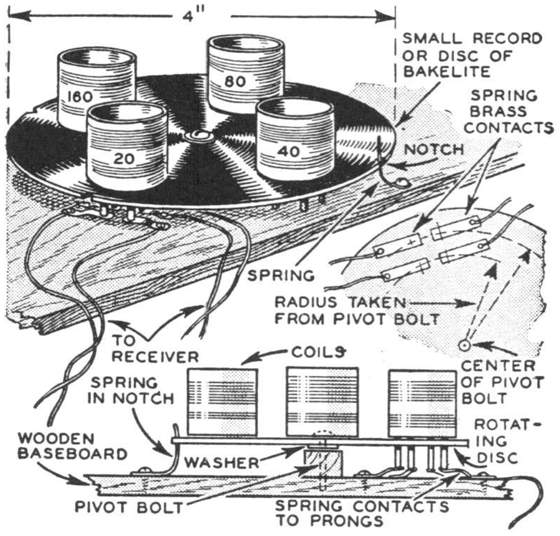


**For quick shift of wave bands, the coils should be mounted on small record, which is in turn mounted to revolve on center block. Prongs then touch brass prongs on baseboard.**

HERE is the germ of an idea. It demonstrates how one using a short wave receiver employing plug-in coils can arrange them so that by the twist of a dial one wave length coil or the other can be put in the circuit instantly.

The usual short wave receiver uses four coils covering the 20-40-80 and 160-meter bands. In this wave band changer, a small toy phonograph record or disk of bakelite is used for the base.

Near the edge, and equidistantly apart, drill four sets of holes for the prongs of the different coils.

Arrange the coils in rotation from 20 meters up, around the edge, and so the corresponding pairs of prongs of each coil will face the same direction. That is, have each pair of large prongs face the inside of the disk towards the center.

Each coil can be fastened solidly to the disk by a tiny angle piece so it will not spring up from the disk.

Now mount the disk on the receiver baseboard by placing a bushing and washer under the disk and running a bolt down through the center hole upon which the disk can rotate. The coil prong ends should be lifted at least ¼ inch above the base.

Next arrange four contact springs made from springy brass upon the base so the four coil prongs from any one coil will come in proper contact with them when the coil is directly over them.

**Contact Springs Made From Brass**

This is suggested in the sketch but the proper shape, etc., will have to be worked out by each builder. Wire these springs to the portions of the circuit to which the coil prongs should go if mounted as usual in a socket.

Lastly, cut notches in the edge of the disk into which a stiff wire catch will spring when the prong contacts are touching the springs under them properly.

When operating properly this coil-changing disk can be instantly rotated and the coil for any band brought into position with the circuit for reception on that desired band.

# Build Midgy Jr. — A Complete Vacation Receiver in Small Cigar Box

*(Continued from page 95)*

to coil up some lengths of fine, flexible wire with clips on one end of each and tuck these inside the cover of the receiver. The extra tuning coils can be carried in the space above the tubes.

For use in a car equipped with spark suppressors, use the car frame as a ground and run the antenna wire out the window to a pole lashed to the side. In a canoe or other boat trail the ground wire in the water and fasten the antenna to the mast or fishpole.

In camp, the well pipe can be used for grounding and an antenna erected in the usual way.

Now that the tubes are lighted you are ready to tune in. Rotate the tuning condenser with the lever, at the same time advancing the variable resistor until a hissing is heard in the phones. This should come in gently. If it plops in with a thud try a higher resistance grid leak (or a different one) or reduce the number of tickler turns on the coil.

When a station is heard the regeneration should either be increased or backed off until the signal is at maximum without the set going into oscillation.

The circuit uses the following parts:

Midget book type balancing condenser,
.000050 variable midget tuning condenser (Pilot or similar),
.0001 fixed condenser, (grid condenser),
.00025 blocking condenser,
7 to 9 megohm grid leak,
8 ohm filament resistor,
50,000 ohm variable regeneration resistor,
Small 3½ to 1 ratio audio transformer,
Two 230 type 2-volt tubes,
Pair of phone tip jacks,
A battery—two cells of flashlight battery, (2⅝x1⅜ inches), 5 cents each,
B battery—Burgess portable 22½-volt B or C battery measuring 3 5/16x2 1/16 x2⅝ inches.

# Single Button Lapel "Mike" Can Be Built Into Watchcase

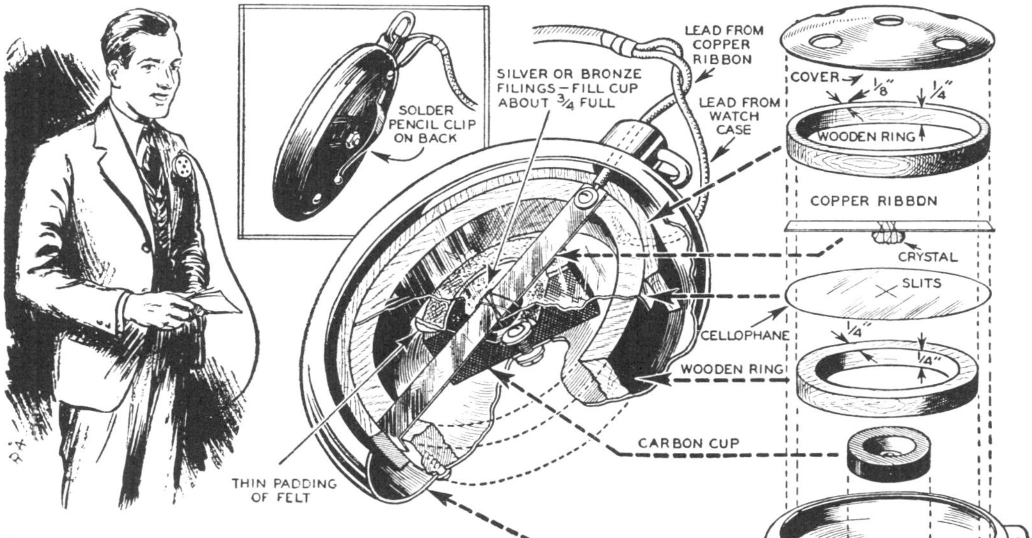


Tiny lapel "mike" works well on low power phone transmitter or public address system. Connect to a 3-volt battery in series with single button "mike" transformer.

A LAPEL microphone which can be fastened to the amateur's coat and carried with him as he moves about is a valuable addition to any "ham" transmitter.

The case from an old dollar watch makes an ideal shell. Replace the crystal with the solid back from a similar watch.

Turn out two concentric wood rings which will fit snugly one inside the other, and just go into the watch case. Fasten the inner ring to the case with short screws.

Cut out the carbon cup from an old dry cell electrode, and glue felt to the rim, to act as a buffer for the diaphragm. Bolt the cup to the case.

Cut a ribbon ¼" wide, of length equal to the diameter of the outer ring, from thin copper foil. Fasten a small piece of detecting crystal to the middle of the strip with fine, thread-like copper wire.

Cut a disc the diameter of the larger wood ring from a sheet of unwrinkled cellophane, and make two short cross slits in the center. Push the crystal through the hole.

Fill the carbon cup ¾ full of silver or bronze filings. Metal filings tend to eliminate the annoying "hiss" of carbon granules.

Place the cellophane diaphragm, with crystal in the filings cup, over the inner ring. Press the outer ring down carefully until the cellophane is well stretched.

Fasten one lead to the ribbon with a washer and screw, the other to the case. Drill about five half inch holes in the front cover, to allow sound to reach diaphragm.

## Rod Antenna Stops Static

WHERE space for a receiving antenna for a short wave set is limited, this sliding brass curtain rod is ideal. The length of the rod is changed to tune the antenna to the station. A variocoupler connected in series with the antenna further tunes the picked up signals for loudest reception. An antenna like this greatly reduces static impulses and other interfering noises.

Though the volume obtained is surprisingly good, it is of course not equal to that to be had with an overhead antenna.

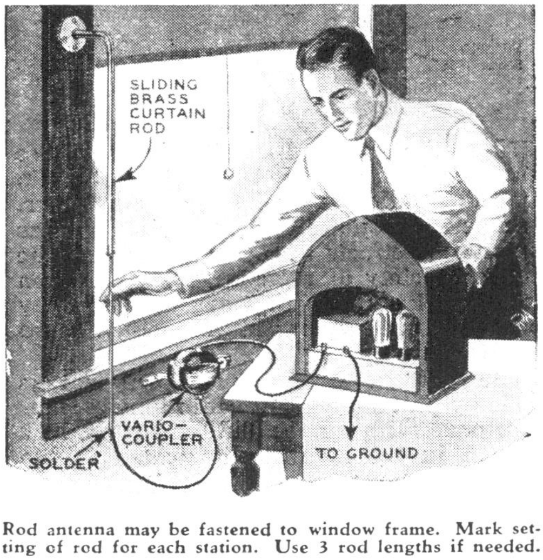


Rod antenna may be fastened to window frame. Mark setting of rod for each station. Use 3 rod lengths if needed.

# SEE RADIO SIGNALS WITH THIS VIBROMETER

Music and voice from radio is produced in sound patterns on the screen of the "vibrometer," as shown in drawing above.

Interior view of the "vibrometer" shows how mirrors are mounted on the 2x4. Beneath is seen the can over which is stretched the rubber holding the pool of mercury. Light beam reflected from mirror fluctuates on screen due to motions of the mercury mirror.

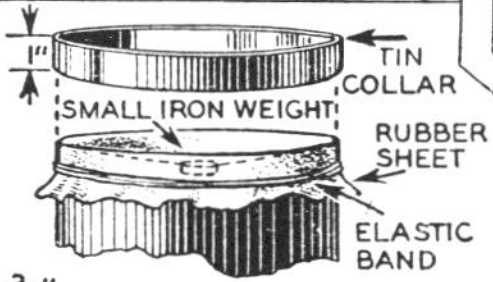


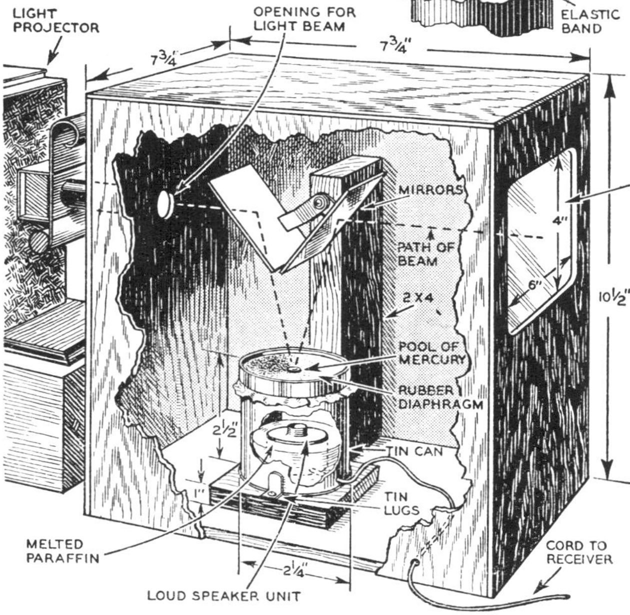


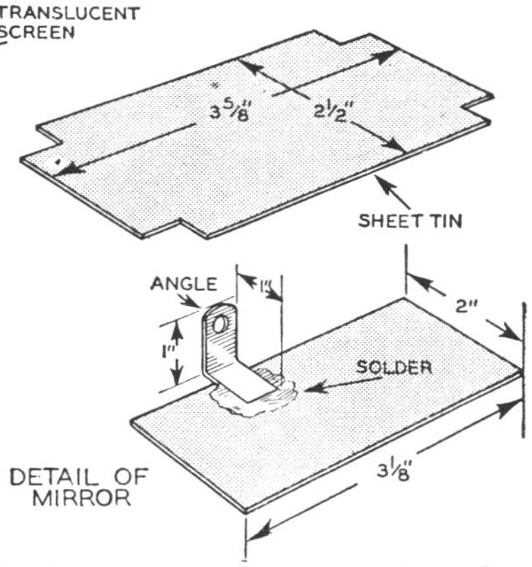


Details of construction are seen here. Considerable care is required in the proper adjustment of mirrors. Beam strikes center of screen.

WITH the simple little instrument here described, a whole new world of radio experimentation is opened to the amateur. On a screen you are enabled actually to see the strange effects of sound waves and the endless variety of patterns, many of them curiously beautiful, that are produced by the incoming music and voice in a radio receiver.

The pattern producing instrument is made by placing a small radio loudspeaker unit, preferably of the Baldwin type, in a tin can in the manner illustrated in the drawings above. A half pint lacquer or paint can is just the right size to accommodate the average small unit.

Two tin lugs are soldered to the can to hold it down to the baseboard. Inasmuch as a dead sound chamber must be produced between the loudspeaker unit and the rubber diaphragm, it will be necessary to pour melted paraffine in the space that exists between the loud speaker unit and the tin can.

The rubber diaphragm must be very thin for best results. The writer used a rubber balloon and strong rubber bands to bind it over the top of the can. The degree of tautness of the rubber diaphragm will depend somewhat upon the power the loudspeaker and the current fed into it. It's

*(Continued on page 99)*

# Good Microphone Stand Easily Made From Music Rack

WHAT transmitting amateur has not longed for a good microphone stand to make the shack look more professional? Most every home possesses one of those old folding metal music racks; it can be easily converted into a nice little mike stand.

Straighten out the bent portion at the top where the rack slips on, or saw it off entirely. Run the saw down lengthwise for about 2″ splitting the hollow shaft. Bend these halves outward until they form a nearly flat, horizontal piece. Heating may be necessary but with such light metal the bending can probably be done without it.

Now get a piece of fairly heavy strip brass about 1″ in width, bend it into a true circle, and solder the ends together. Its diameter can be anywhere between 6″ and 8″. Clamp the soldered place over the bent portion of the stand, drill small holes through stand end and ring ends, and bolt in position vertically with two small bolts.

Drill four small holes equi-distantly through the ring and insert four small eyebolts with the eyes on the inside of the ring. These are for supporting the microphone.

Instead of metal springs, use pieces of heavy rubber bands inserted first through a ring eye and then through one of the eyes on the mike. This will support the mike centrally inside the ring and the rubber will do away with microphonic noises that often occur when springs are used.

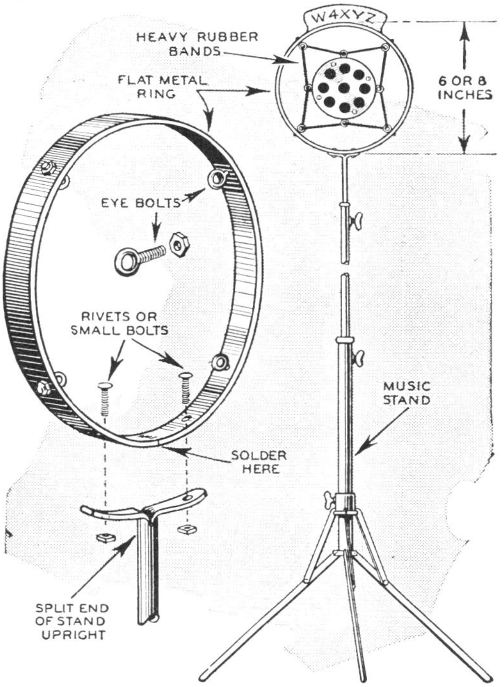


## "Vibrometer" Makes Music Visible

*(Continued from page 98)*

mainly a "cut and try" job.

Inasmuch as a small pool of mercury is placed on the rubber diaphragm, it will be necessary to create just the slightest depression in the center. This is done by hanging a small weight from the center.

The cabinet proper is made of plywood with a ¾-inch soft pine base. The window in the front is cut out on a band or jig saw and provided with ground glass. The images of the sound waves are created by the light reflected by a series of mirrors, the mercury itself acting in the capacity of a mirror. These mirrors have tin backs of the size and shape illustrated and are attached to the posts with angles.

The source of light may come from a small magic lantern or the experimenter may easily build himself a small light projector. The more powerful the source of light the better.

It will be necessary to do a little adjusting to make this device perform perfectly. In arranging the mirror at the correct angle, it should always be borne in mind that the angle of incidence is always equal to the angle of reflection.

## Make Your Own Little Fuses

THESE three holders for small fuses will save many a meter and radio tube.

Mount two binding posts about 2″ apart on a bakelite base. Under one fasten a piece of spring brass wire. Tie a short piece of fuse wire between the spring and the other binding post, leaving a slight tension in the fuse wire.

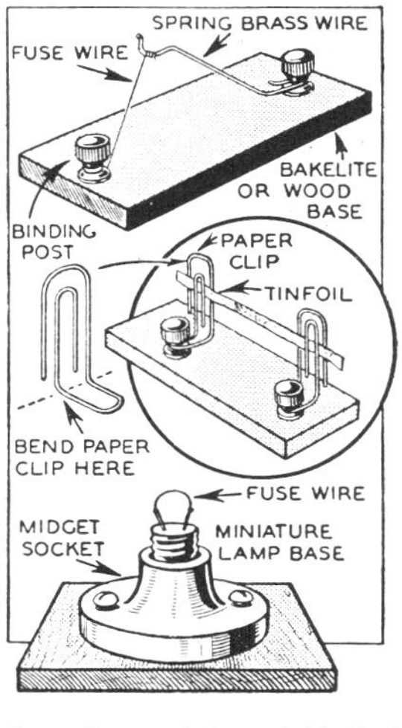


If no fuse wire is handy a strip of tinfoil 2″ by ⅛″ taken from a cigarette pack will do for a fuse. Bend two paper clips at right angles and fasten them under the binding posts. Insert foil strips between the clips.

Mount a miniature lamp socket on a base. Break the glass from old flashlight bulbs, and loop pieces of fuse wire across the filament leads. Very thin tinfoil strips may also be used here, just as was done in the above fuses.

# PRINT DISTINCTIVE Personal QSL CARDS

The finished QSL card will appear as a white design on black card. White ink can be used in filling out card.

THE array of QSL cards on the wall of any amateur's shack is seldom anything to attract the eye. Except for color, the cards are all just about the same.

Individualized cards in artistic designs can easily be made by an inexpensive photographic process. If you are a bit artistically inclined, you can draw your own design; otherwise get some artist friend to draw up your ideas.

First lay out your card to exact size in pencil on drawing paper, changing things around until you are satisfied with the design. Lay a piece of architect's tracing linen or transparent paper on the pencil sketch, and copy the drawing in black ink.

In a dark room lay the design ink-side down against the glass of a printing frame, place the emulsion side of a photo post card down on the linen, and clamp in the printing frame pad. Expose, develop, and print just as is done in printing from any negative. The design will appear in white on a black card, so white ink should be used in writing on the cards.

If desired the original pen and ink drawing may be made oversize, and a postcard size photograph taken of it. Cards can then be printed from the negative in the usual manner. The design will here appear in black on the white card.

## Killing Generator Interference

THERE are still many farm electric plants, both 32 v. and 110 v., which can cause bad radio interference. A home made filter costs very little and cuts out all radio interference from this source.

Mount two 1" to 1½" wood dowels each about one foot long on a wood base. Wind on each about 230 turns of insulated No. 6 or No. 8 copper wire. Wind the turns close and tight, fastening the ends firmly to the dowels. Be sure that both coils are wound in the same direction.

Break the two wires running from the generator. Connect the generator wires to the bottom ends of the two choke coils. Connect the top ends to the battery wires.

Get two paper condensers, of 3 to 4 mfd. capacity, being sure that both are alike. Connect them in series across the bottom ends of the choke coils, and connect the mid point to some very good ground.

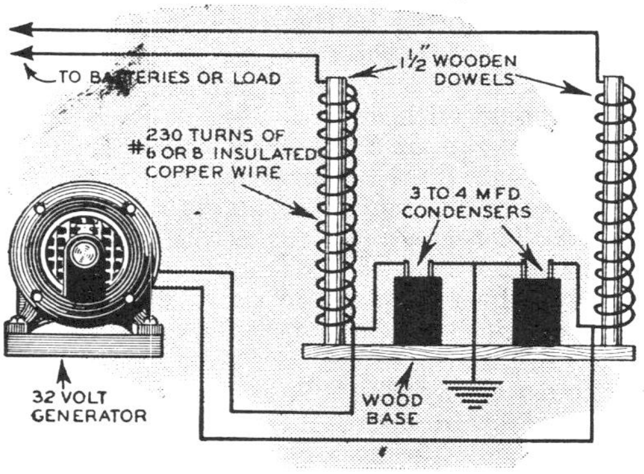

# Powerful CRYSTAL SET Brings In Distance

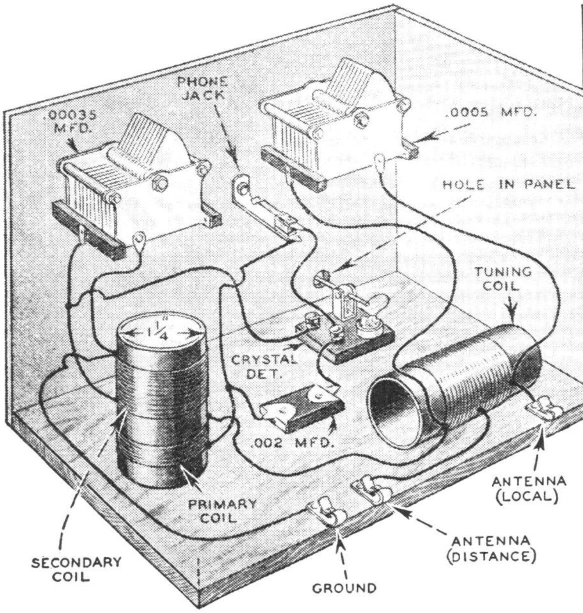


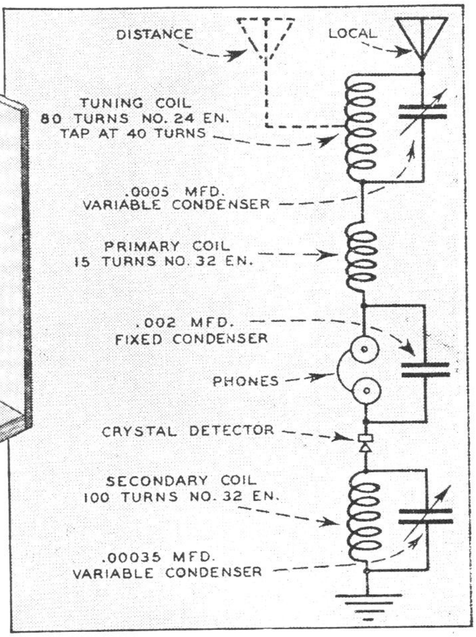


**Suggested arrangement of parts for super-powerful crystal receiver is shown above. White pine baseboard is 1" thick; use hardwood for rest of cabinet. Bakelite or metal front panel may be used if desired; in this case, ground panel, and insulate phone jack and .0005 condenser from it. Circuit diagram in simplified form as shown. Use mica condenser across phones.**

CRYSTAL sets are once more coming into favor with radio listeners, principally because the great increases in power made within the past few years by the larger broadcasters make distant reception possible.

This highly selective crystal set has brought in stations as far as 1000 miles away; a log of the stations it has picked up includes an amazingly high percentage of the high-power stations in this country. With sensitive and high-quality headphones, a good aerial, and a good ground, you, too, can enjoy the thrills of crystal-clear DX with not a single battery to worry about.

The antenna tuning coil is wound on one coil form, with a center tap to give added sensitivity when distant reception is desired. Place across this an .0005 mfd. variable condenser, which generally will have about 21 large size plates. If you do not happen to have such a large condenser at hand, simply connect two smaller units in parallel, set one with the plates completely meshed together for maximum capacity, and tune with the other.

Both primary and secondary coils are wound on the same form. The size of wire used for any one of the three coils is not extremely important, nor is the insulation, so use whatever wire you happen to have at hand.

The tuning condenser across the secondary coil should be a standard broadcast band condenser of .00035 mfd. size. This may be obtained from any old battery radio set.

Use the most sensitive crystal you can get in this radio—better still, order two or three different kinds, and give each a tryout. Great differences in sensitivity will be found even in crystals which appear exactly alike. Remember that heat in any degree will decrease the sensitivity of your crystal; even the small amount of heat set up when currents from a powerful local station are passing through the crystal may produce local heat which will cause a gradual reduction in volume. Reduce the volume on local stations by turning down the tuning control a bit, to prevent damage to a good crystal.

A phone jack with plug is used in order that the phones may be easily removed when needed for other purposes.

A very attractive cabinet can be made for this set from ¼" hardwood. If carefully sanded and varnished or lacquered, your set can be made just as modernistic and attractive in appearance as some of the midget electric sets now on the market.

To permit easy adjustment of the crystal without reaching into the set, a small hole is cut in the front panel. The crystal is mounted so the catwhisker handle projects slightly out from this hole.

# Simple WHEATSTONE BRIDGE for RADIO Experimenters

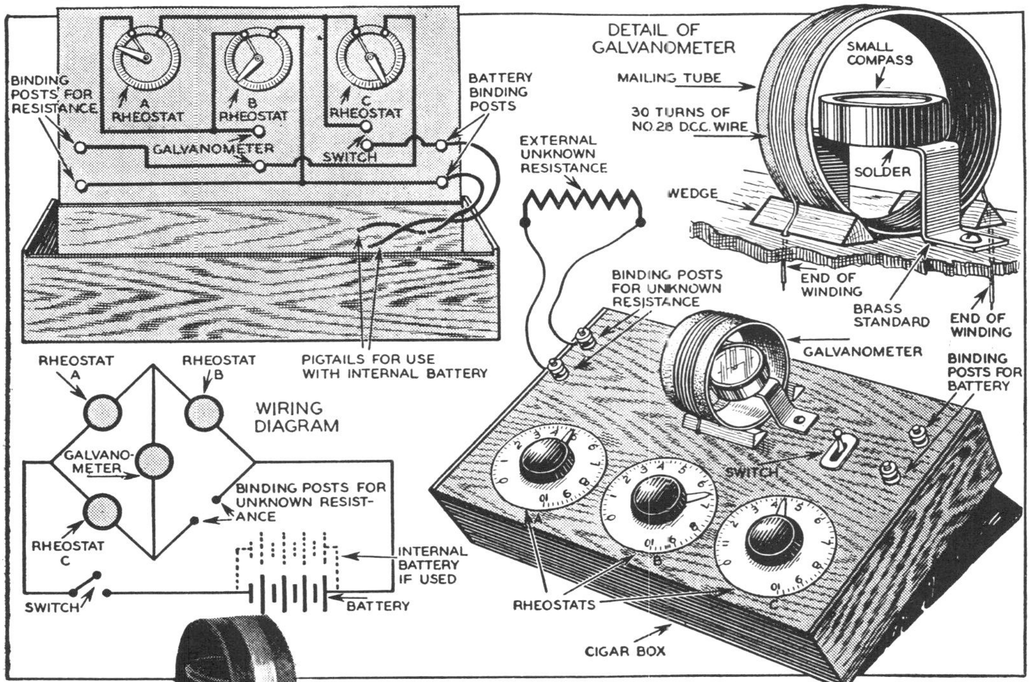


Follow these drawings to wire up the galvanometer. The unknown resistance attaches to the left binding posts, the battery to posts at right. Note wedges holding coil.

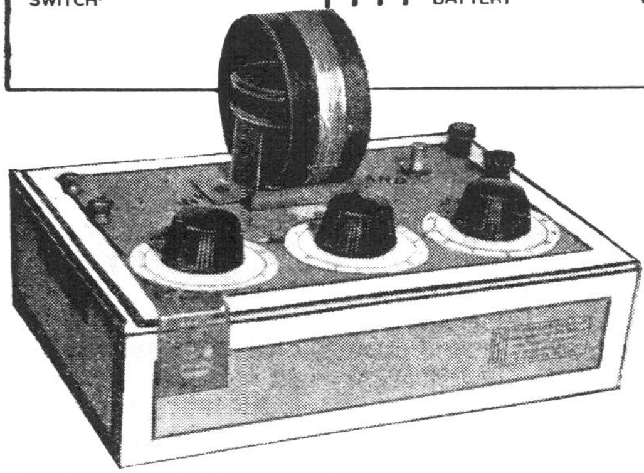

The wheatstone bridge assembled ready to determine unknown resistances. The three rheostat dials are adjusted in the determining process, the galvanometer telling the story.

A WHEATSTONE bridge is one of the standard pieces of electrical apparatus for laboratory use. While a manufactured bridge of extreme sensitivity is somewhat costly, the electrical and radio experimenter can construct his own at very small cost.

This one will do very nicely to measure unknown resistances, find unknown condenser capacities and also inductance values. The one shown will handle up to 1000 ohms as unknown resistance but by substituting larger known resistances, the value of the bridge can be raised much higher.

First place the rheostats in the box lid with the knobs projecting through on the top side and near the front edge. From left to right letter them A, B and C. Then calibrate them by closing all three up so no resistance is included. Draw an arc around the knob and divide it into ten equal divisions of one ohm each.

Then divide each ohm into ten subdivisions of one-tenth ohm each. This will give you 100 points under each knob for accurate readings. A paper scale can be drawn and pasted under each knob or the scale can be inked directly on the lid itself.

The galvanometer is easily made. Cut off a section of 3-inch mailing tube about 1½ inches wide and wind on about 30 turns of No. 28 or No. 30 DCC wire as shown in the top drawing. "Sew" the ends of the turns into the cardboard of the tube after pulling the windings tight and then cover the latter with a coat or two of shellac to bind them. Place the coil on the lid back of rheostat B fastening it with a couple of little bolts through the lid.

Then solder the base of the compass to a piece of stiff brass about ½ inch wide and bent in the form shown in detail to act as a support. Fasten this support to the lid at right angles to the coil and so the compass sits about in the center of the coil. The details of the galvanometer and its finished appearance are clearly shown in the sketch.

*(Continued on page 103)*

# Power Tube Panel Mounting Made From Piece of Conduit

HERE'S an ideal socket mounting for tubes on public address amplifiers of amateur short wave transmitters. A short piece of 1½ in. malleable iron conduit is heated and placed in a vise. The jaws are then closed rapidly to flatten the end of the pipe. This flattened portion is then drilled for socket mounting and wire holes.

The nuts and washers ordinarily used on the conduit are employed to secure the pipe firmly to the front of the panel, as illustrated in the drawing at the right. Wires leading out of the rear are connected into the circuit.

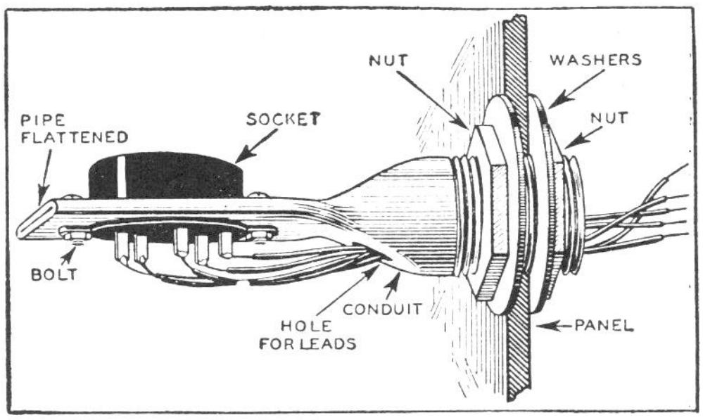


Mount the tube socket on flattened end of the conduit, and secure round end on front of panel with nuts and washers. Panel mount bakelite sockets are easiest to mount here, but ordinary basemount socket will do if large conduit is used.

## Remove Shields for More "DX"

REMOVING the shields from screen grid tubes in your set will often give it regenerative qualities, bringing in many more distant stations. Tuning will be far more sensitive than before.

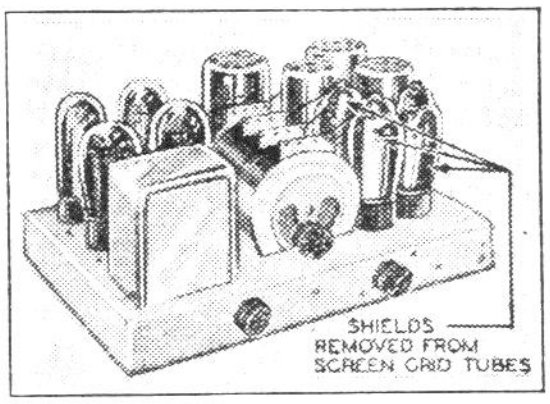


Remove the screen grid tube shields, then replace grid leads.

## Bridge for Radio Experimenters

*(Continued from page 102)*

In wiring, use No. 18 insulated wire to keep all undue resistances out of the circuit, which is given in the drawings.

Finish off the box and give it a coat of stain or paint and you will have a nice looking instrument.

The principle of the Wheatstone Bridge is that four resistances are in a circuit as shown in the diagram; one of them being unknown. Switch on the current and, if all resistances are alike, no current will flow through the galvanometer and its needle will remain quiet.

If the unknown resistance varies from the others then the rheostats A, B and C will have to be adjusted until the needle shows no deflection. Knowing the values of A, B and C you can then use the formula A over B equals X (unknown) over C.

Taking the scale readings of A, B and C it will be easy to determine the resistance of the unknown X. Any text book will give you the various methods of determining unknown inductance and capacity values too lengthy for this article.

This instrument will be found especially valuable in winding filament circuit resistors or making small, accurate resistors from high resistance wire.

## Making Longer Plug-in Coils

THE only disadvantage encountered in using old radio tube bases for plug-in coils on short wave sets is that for the higher wavelengths the bases will not hold the necessary number of turns.

By mounting two or more bases end to end very neat coils can be made.

One of the coils is wound near the top of a tube base, and holes drilled in the base to lead the wires inside. These coil leads are brought through two of the tube prongs and soldered in the usual manner.

On a second tube base the coil is wound so that its lowest turn is near the bottom of the base. Leads are brought inside and soldered to two pieces of stiff wire which have been brought in through the prongs. The stiff wire is soldered to the prongs, and then threaded through the two empty prongs of the other tube base. The two bases are brought close together and the stiff wires soldered to the bottom prongs.

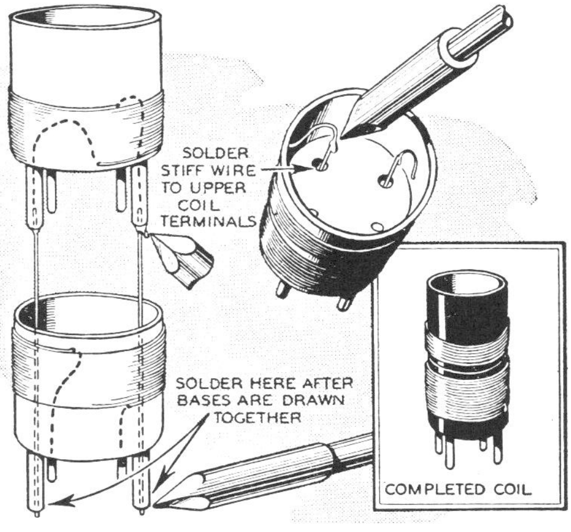


Wind coil on lower tube base first, anchor ends, then fasten two bases together as shown above to obtain room for second coil. Lacquer or shellac coil for permanence.

# Hear All-Wave Programs With

**Though no plug-in coils are used, programs down to 35 meters can be heard with speaker volume on this powerful long and short wave portable receiver.**

Set your toolbox portable anywhere, connect up antenna and ground leads, flip the switch, and take your pick of music, police calls, or amateur short wave broadcasts o'er the continent. Switch for long or short antenna is on back cover.

SIMPLY connect up the aerial and ground wires, and this powerful little receiver is ready to pick up stations anywhere in this country, on any band from 35 to about 370 meters, with good volume on the built-in midget loudspeaker. Take it anywhere—on a canoe, cruiser, or along to camp this summer—it doesn't take much of an aerial to pick up stations on the broadcast band hundreds of miles away.

An ingeniously wound coil covers four different bands simply by changing the setting on a four point selector switch. There is none of the nuisance of carrying along plug in coils for the different short wave bands, and opening up the set each time you wish to change bands.

Everything is built into the single tool box cabinet. Light weight B and C batteries give unusually long service, since the current drain with the circuit used is very low. Two dry cells supply filament power.

Rain or snow have no threats for this sturdily built receiver. Its dependability makes it the ideal portable set for mining expeditions, explorers, forest patrol rangers, or even for interference finding. A

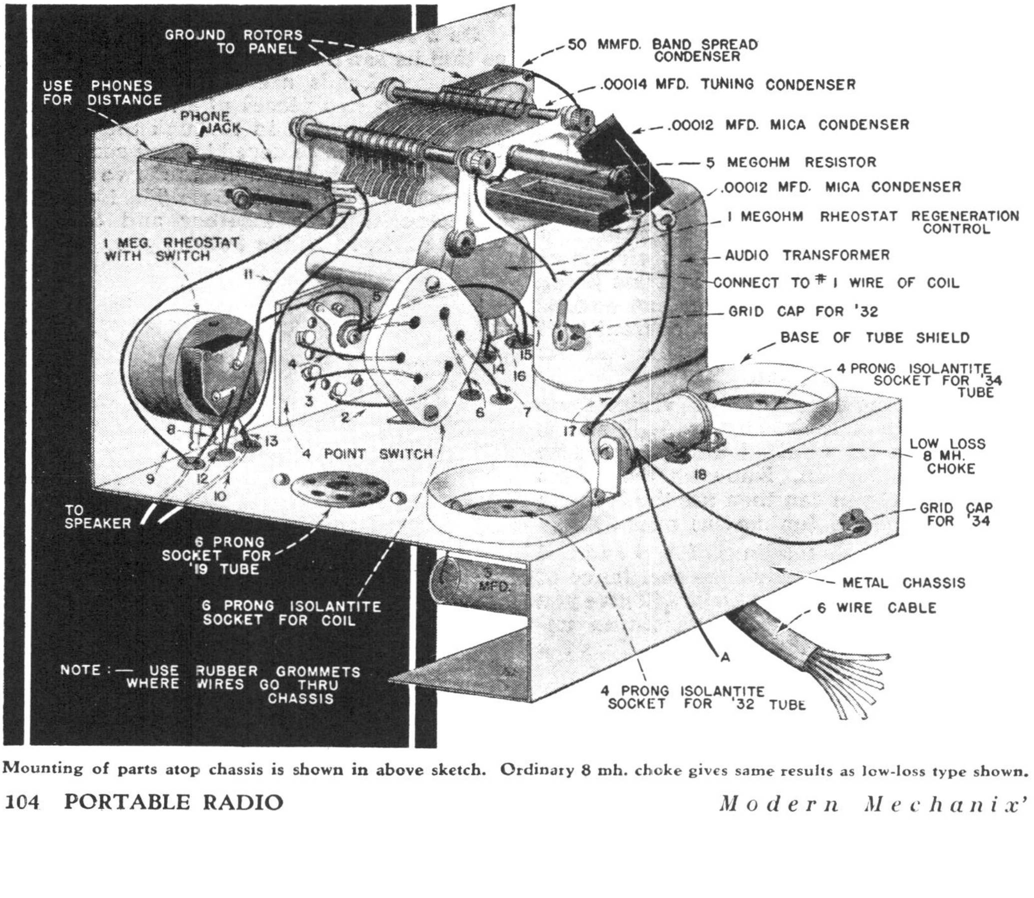


Mounting of parts atop chassis is shown in above sketch. Ordinary 8 mh. choke gives same results as low-loss type shown.

# 3-TUBE TOOL BOX Portable

*by* P. M. OHLINGER

phone jack is provided to permit headphone reception of distant short wave stations. When phones are used, the speaker is automatically shut off.

Though only three tubes are used in this set, the type '19 dual purpose tube raises the efficiency to that of a four tube receiver. Electrically the circuit is a '34 screen grid untuned r.f. stage coupled by choke and condenser to a '32 screen grid regenerative detector. Resistance coupling is used between the detector and the first a. f. stage, which is one half of the '19 tube. Transformer coupling is used between first and second audio stages, the output passing through the phone jack to the speaker.

A tone control is just another of the many features found in this circuit. It may be used either for selecting the tone desired or for cutting down background noises on local reception.

A 16″ by 7½″ by 7¼″ 22-gauge steel tool case is used as a cabinet. It presents a fine appearance and takes very little machine work.

Drill out the rivets that hold the handle to the lid. The box is stood on end and the handle fastened to the top. Bolt a sponge rubber pad to the bottom of the box to take up shock and vibration. Drill holes for the antenna, ground posts and volt-meter on the rear door. The front panel of the box must be drilled or cut out for the speaker opening

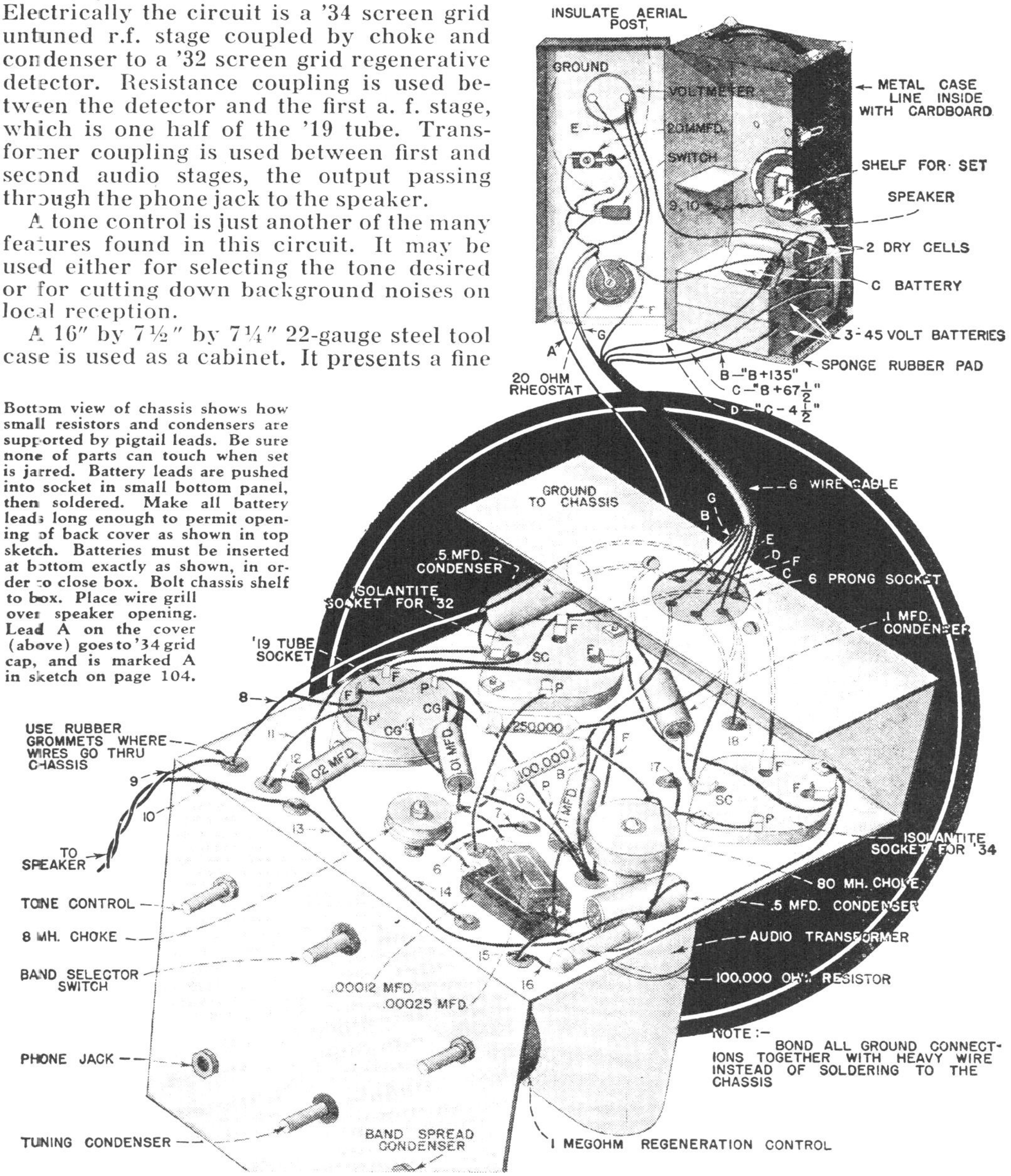


Bottom view of chassis shows how small resistors and condensers are supported by pigtail leads. Be sure none of parts can touch when set is jarred. Battery leads are pushed into socket in small bottom panel, then soldered. Make all battery leads long enough to permit opening of back cover as shown in top sketch. Batteries must be inserted at bottom exactly as shown, in order to close box. Bolt chassis shelf to box. Place wire grill over speaker opening. Lead A on the cover (above) goes to '34 grid cap, and is marked A in sketch on page 104.

## *Chassis of Portable Set Slides Out Easily When the Dials Are Removed*

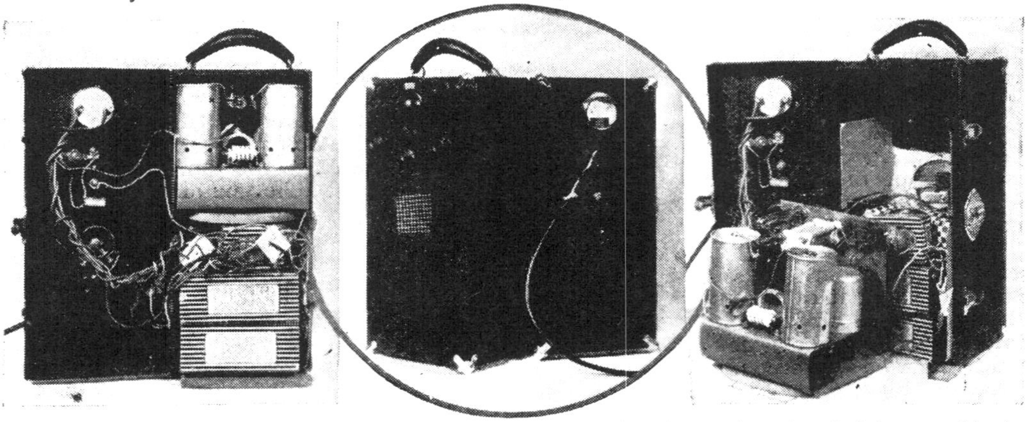

Three views of completed set. To slide out chassis, loosen screw on tuning dial coupling, take off all front panel knobs.

and for the controls. The five inch magnetic speaker will fit in very easily.

The inside of the box should be lined with cardboard to prevent battery shorts and to make acoustics better.

A vernier dial is fastened to the control panel by its extended bolts. Then a bushing with a ¼-inch shaft extender couples the dial to the tuning condenser. This shaft and coupler may be made from an old bakelite dial. Break the dial and remove the bushing. Cut off a piece of ¼-inch shaft from an old control or condenser. Solder one end of this shaft into the bushing so that the set screw will be free to engage the condenser shaft. The chassis may then be removed without taking off the tuning dial.

If an electralloy panel 7x18 inches long is used, there will be a three inch strip left over. This may be used to make two brackets which are bolted to the inside of the box about eight inches from the bottom, to support the chassis. A sponge rubber base is placed between the chassis and these sup-

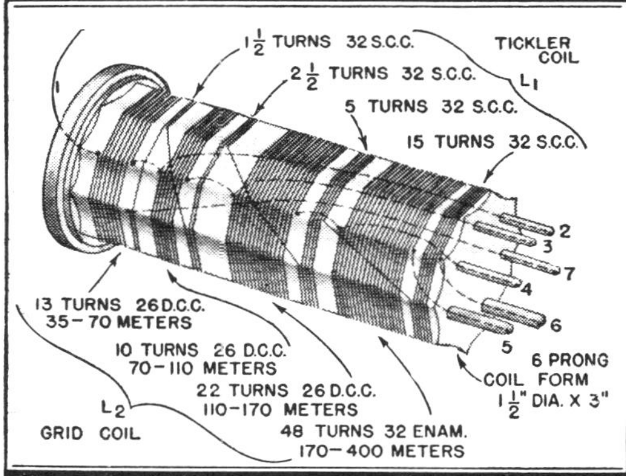


Wind coil exactly as shown, spacing grid windings 5/16" apart. Circuit diagram is given below. Filament rheostat is adjusted to control volume and regeneration. Filament meter should never be allowed to go higher than 2 volts. Use low-priced 3-volt D. C. meter. Aerial binding post is on back cover. Ground post, on cover, is not insulated from tool box.

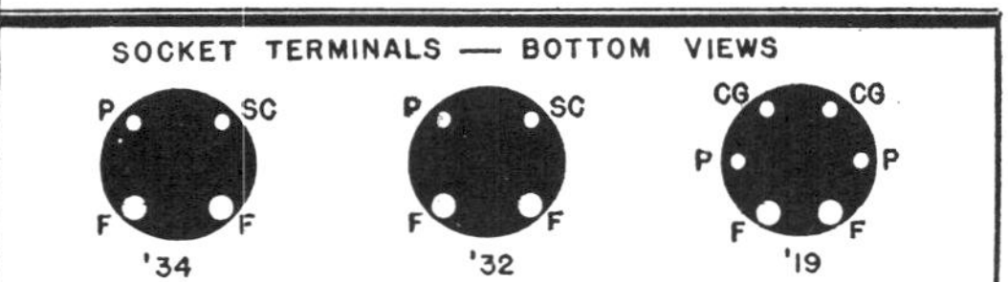


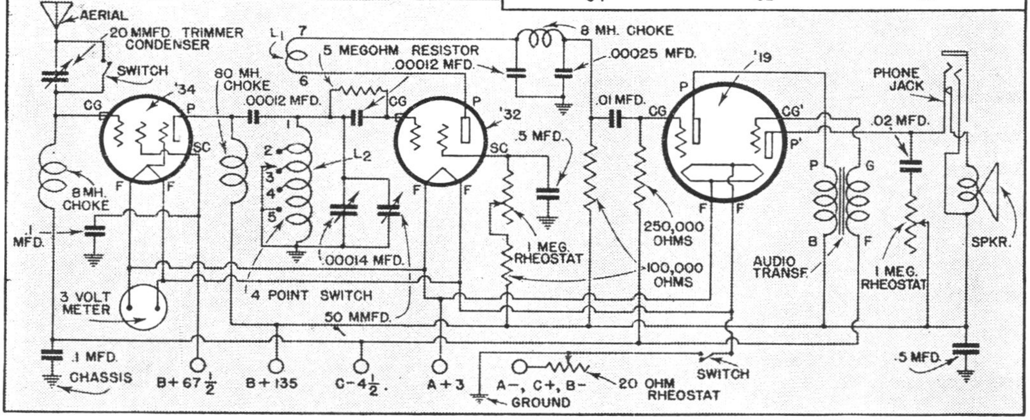

# Heat LEAD and SULPHUR to Make RADIO CRYSTALS

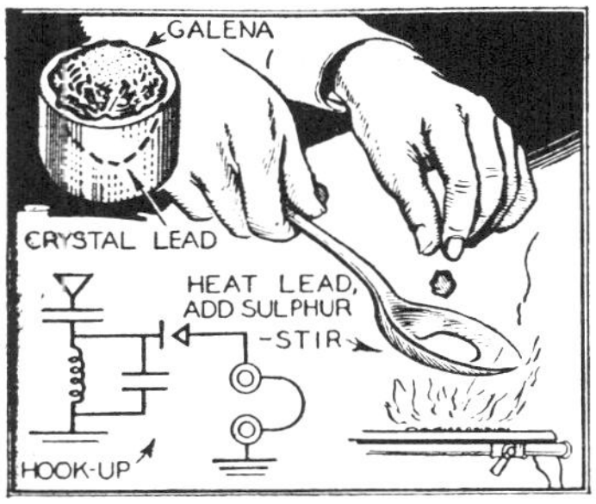


CRYSTALS of the galena type, used extensively a few years back for crystal radio receivers, can be made in a few minutes by radio experimenters. Only a small piece of lead and about a gram of sulphur are needed. Place a piece of lead about the size of a pea in an old teaspoon, heat until the lead melts, then add one gram of sulphur and stir the mixture with a piece of wire. Allow to cool, and use the resulting black metal exactly as you would a crystal.—C. K. Jones.

## Ford Magneto Terminals Make Good Insulators

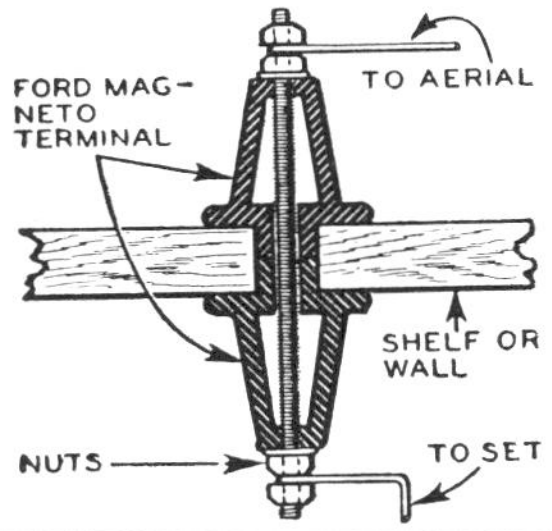


MAGNETO terminals from the old model T Ford may be used to advantage as stand-off insulators by short wave enthusiasts. Remove all dirt and grease with gasoline. When using them to bring an aerial wire through a wall drill a hole the size of the terminal base, and bolt two terminals together as shown, one on each side of the wall.

## Hear All-Wave Programs With 3-Tube Tool Box Portable

porting brackets. Two small pieces of sponge rubber will dampen tube vibrations if placed in the tops of the shield cans.

Bend down the two flanges that formerly held the tray in the box. Form the chassis from a strip of electralloy 7x15 inches long. This must be cut down to 6½ inches in width, and is easily done. Take a rule, mark off ½ inch and follow this line with a sharp pointed instrument until a deep cut results.

Place the panel in a vise and bend back and forth. It will break off cleanly where the line has been cut. Bend this panel to dimensions given and drill holes.

The battery cable fastens to a six prong wafer socket which is fastened on the bottom chassis flange. Rubber grommets or bakelite bushings of 5/16" or ½" size are used where wires pass through metal.

Use moisture proof hook-up wire and solder joints well, making resistors and condensers vibration proof. Wire in filament circuits first. The filament switch is on the shaft of the tone control.

The high impedance R. F. choke may be mounted on a stand-off insulator or soldered directly on the tube socket plate terminal. Study the connections to the six prong "19" tube carefully before wiring.

A means of cheap insurance against tube burn-outs is used in this set by wiring a flashlight bulb in series in the B lead. Any B battery short circuit will blow this bulb and not harm the tubes in the least. This fuse may save the cost of several sets of tubes. All B batteries may be disconnected by unscrewing the flashlight bulb.

Negative wiring and grounds are soldered to any machine bolts which protrude through the underside of the chassis; these are all bonded together by soldering a bus wire to them. This gives a well defined path for return of R. F. currents.

The coil form is a 6 prong low loss type made for short wave work. It fits into a six prong Isolantite socket, from which connections are made to a four contact selector switch mounted on the front panel. The grid winding is tapped in four places to allow selection of four different bands. These taps are brought to four prongs on the coil form. Though the grid winding is continuous, it is in effect broken when the grounded switch arm contacts the different taps and shorts unused coil portions to ground.

The two remaining coil prongs are for the tickler winding; this is in one winding but distributed with different turns for different coil sections. Grid coil windings are spaced five-sixteenths of an inch apart. Tickler windings must be wound and spaced between by these experiments. If oscillation cannot be obtained over a part of a certain band, push grid and tickler coils for that band closer together.

The coil may be removed and windings changed without disturbing any of the other parts or wiring.

BLUEPRINTS for this three-tube toolbox portable receiver, showing clearly every detail of the sketches on these pages, are available to radio experimenters at 50 cents postpaid. These blue-line prints, on tough white paper, contain in addition a complete detailed list of the parts needed to build the set. Modern Mechanix Pub. Co., 529 S. 7th, Minneapolis, Minn.

# A TIME-SAVING SWITCHBOARD for the Radio Bench

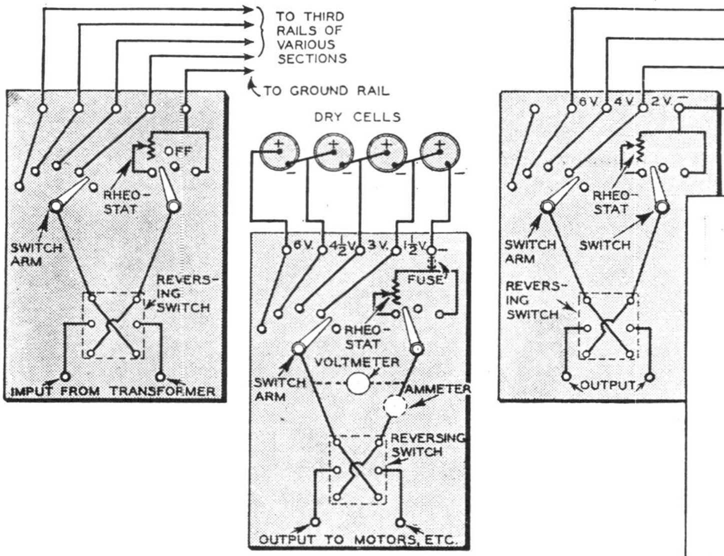

In wiring up your switchboard select the circuit above that best suits your needs. Voltmeter and ammeter are optional, but they will prove extremely serviceable for plating jobs. Reversing switch shifts current from positive to negative instantly.

## Neon Tube for Night Signaling

BOY Scouts who like to do night signaling will find good use for this little neon blinker set. First you'll have to procure a neon tube and a key. Mount the tube in a reflector as illustrated below and hook up the key as in the drawing. Plug the light in a 110 volt outlet and bang away at the key. One advantage of the tube is that the flash is instantaneous. The light comes to full glow instantaneously and shuts off instantly, making for unusual distinction in night signaling. For a key, any old telegraph signaling key will do duty excellently.

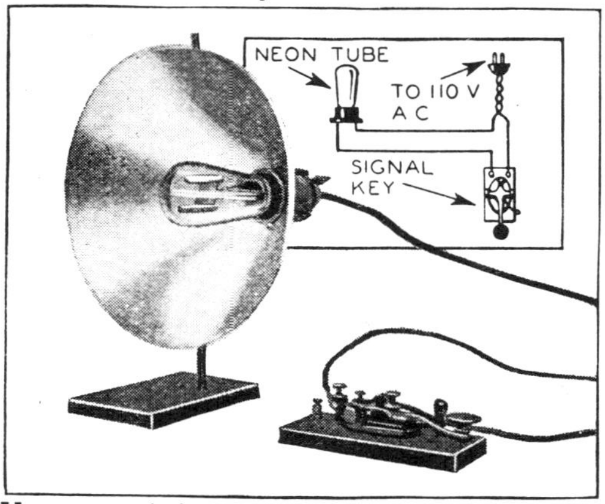


Mount neon tube in a reflector as illustrated in photo. Tube is hooked in series with key and 110-volt bayonet plug.

ARE you one of the wire twisting fraternity who has grown tired of perpetually changing wires every time you decide to reverse the current or change the voltage to one of your pet motors or generators? If you've wanted a switchboard to do the work here's the answer to your prayers.

This switchboard will work fine with a storage battery, dry cells, or a toy transformer. It will allow you to change the voltage in several steps, reverse the current, switch it off, use the rheostat to dim the lights, or switch the current to different circuits. The apparatus will carry in excess of 500 watts.

An old radio panel may be put into service, and if you like an ammeter and voltmeter may be added for use in electroplating and other experiments. For safety's sake a tubular fuse of the automotive variety may be added to prevent overloads.

Mount all the switcharms, switchpoints, and other equipment as suits your convenience. Then wire the board according to the accompanying diagrams. Twisted lamp cord will easily handle the current.

The panel may be used to deliver different voltages, or it may be used to switch the current to the various isolated blocks of a model railway system. In order to reverse the current it is only necessary to reverse the current from the top to the bottom pole.

# Multiple Antenna System Made From Telephone Wires

CONFRONTED with the laborious job of providing four different antennae for his four different types of receiving sets the writer took steps to work out a single antenna system for the whole works.

The telephone service entered the building through the basement wall through a lead cable as shown in the accompanying sketch. On the basement wall was mounted the lightning arrester. This consisted of a slate block on which were two resistors connected at one end to the entering double wire. At the other end were attached the wires leading to the phone and across these ends was a vacuum gap. A third element between the gap pieces was grounded. Now comes the simplicity of the antenna stunt.

From one of the gap ends of a resistor a lead-in wire was run up through the floor to the room where the receivers were located. A fixed condenser of .0005 capacity was inserted in the lead-in in series. From this condenser a short piece of wire was used to connect to one end of a piece of brass rod about 4 ft. long. Four spring clips were soldered to the rod at equal intervals and the lead-ins to the four receivers clipped in.

With this outfit all four receivers can be operated at the same time on different frequencies. Long, short, and medium waves can be received simultaneously without one receiver affecting the reception of the other.

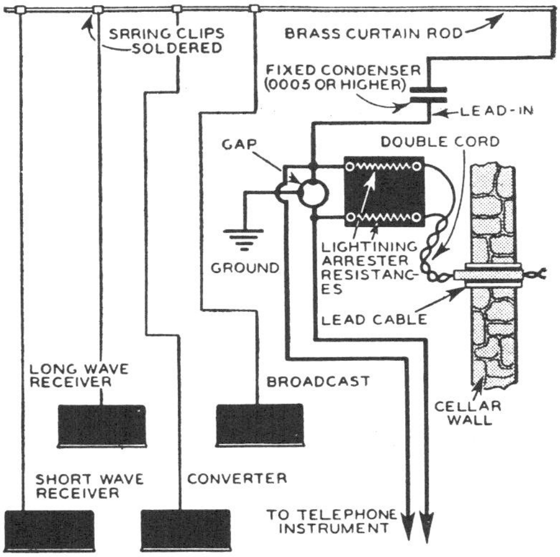


Four receivers are hooked to the brass rod by means of clips at regular intervals. Other sides of receivers are grounded.

## Emergency Choke for B Eliminator

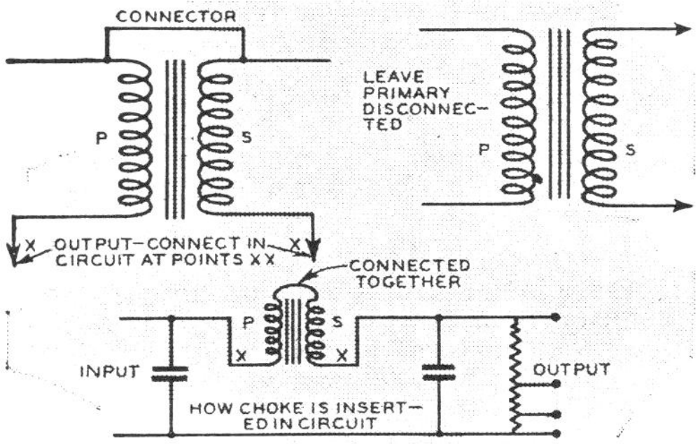


Primary and secondary of transformer are hooked together on one side while other side hooks into the filter circuit.

IF YOU have an old style B eliminator that has a burned out filter choke you can put it into commission again with the aid of a good output transformer made for speaker use.

The secondary can be used alone as a choke or the primary and secondary can be tied together as suggested.

Insert it as shown in the B plus line and use at least a 4 mfd. condenser each side. For an eliminator delivering *not over* 50 mils, this substitute will be found very good in an emergency.

## Antenna for Portable Receivers

HERE is an antenna that is just the thing for use when traveling with a portable receiver or when camping.

The reel is made by using two tin can covers for flanges, a spool in between them and a pivot bolt on a stick. Saw off the flanges from a large spool and center them between the covers, assembling the parts with screws, as shown.

Drill a hole through the exact center of each cover and run a bolt through this bearing hole. Fasten the end of the bolt through a stick for a handle as shown.

Fasten the antenna end to the spool, wind it up on the reel, and to the other end attach an insulator and wire hook for fastening to a limb or other suitable object.

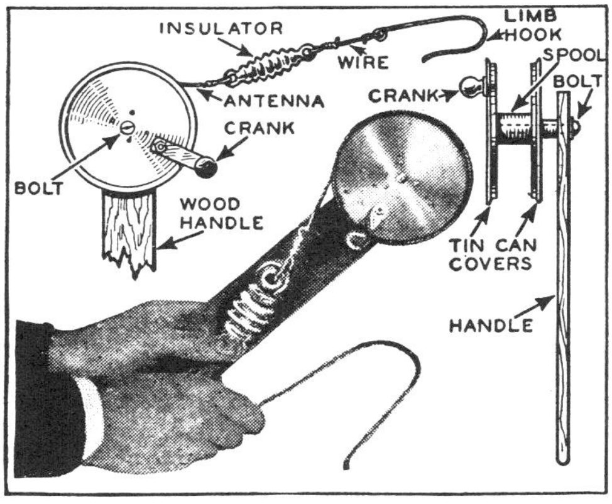


To string up the antenna, attach the hook to a tree limb or other object and unwind the aerial wire from the reel.

# Build Selenium Electric Eye

At left: Photo-electric cell opens a house door when anyone approaches. Light beam is intercepted, and cell operates relay to door opening mechanism. At top: Completed selenium photo-electric cells, and the commercial selenium from which they are made. Below: Cell shows current of 4 milliamperes at 45 volts in dim light, making resistance of 11,000 ohms.

*by* JUSTIN RAU

HOME electrical experimenters can find hundreds of uses for this easily built selenium type photo-electric cell. It is a really practical little "electric eye," and will operate an ordinary primary relay directly without the use of vacuum tube amplifiers.

With this unit any sort of electrical apparatus can be turned on or off at nightfall; garage doors may be opened as a car enters the driveway; the kitchen door can be rigged up to swing open when mother passes the icebox, or it can be put to any other use which depends on a change in light intensity for its operation.

The necessary materials are readily available; many of the items may be found in the experimenter's workshop. Enough selenium for fifty cells can be had from a chemical supply house for about fifty cents.

### Slate Slab Used for Base

A slab of slate will be needed for the base of the cells. Quarter inch switchboard or blackboard slate serves nicely, though a slate shingle may be used if surface irregularities are ground away.

The slate blanks may be cut out with a hacksaw, and the binding post holes, at both ends of the slab, drilled with a hand-drill. Any deep irregularities on the face of the slate should be removed with medium sandpaper by placing the abrasive face-up on a flat surface and rubbing the slab over it. The surface should then be finished in the same way on a glass plate sprinkled with pumice stone and water. Water should be added from time to time during the polishing process. The abrasive gradually breaks down, giving a smooth texture to the surface.

### Graphite Applied to Surface

The surface is made to conduct a current by applying a heavy coating of graphite. A lump of natural graphite is best here, although the lead from a very soft pencil will do. It is well to rub the graphite into the pores of the slate with a bit of soft paper before applying a second coat.

A zig-zag grid should then be cut through the graphite coating, dividing the slab into two electrically separated parts. The grid is formed by scratching, with a sharp knife or razor blade, an odd number of lines across the face of the slab, leaving a border of about a quarter inch at either edge. The first and last line are each extended through one of these borders, so that by connecting alternate lines along the edges the two terminals may be electrically separated. By using a small try-square for ruling the lines parallel, from sixteen to twenty lines to the inch can be cut without difficulty.

When the grid is completed, it should be tested for short circuits. If a low resistance

# to Open GARAGE Door

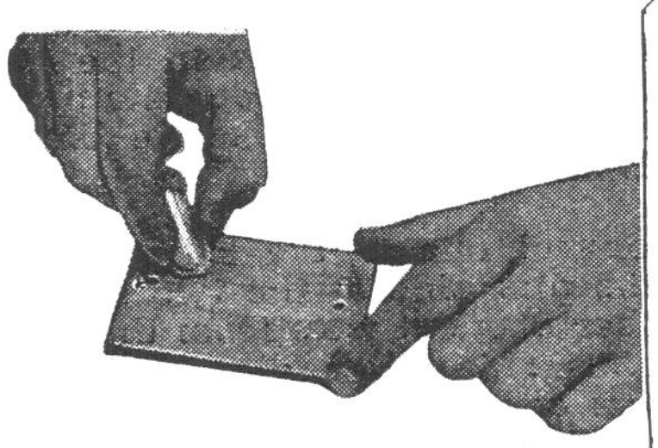

A lump of soft graphite or a soft lead pencil is rubbed evenly over the polished slate surface. Two coats will be needed.

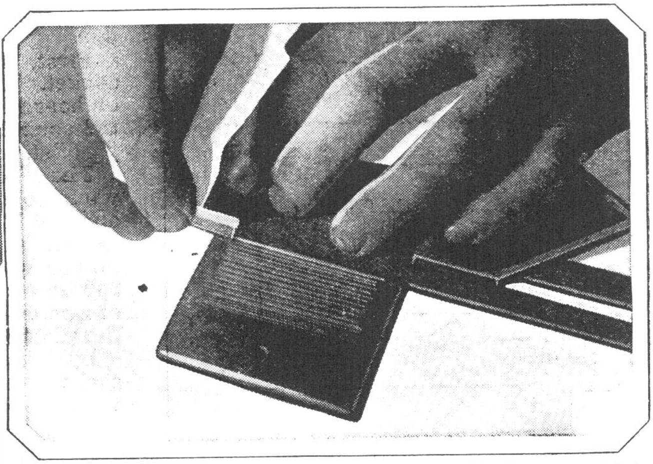

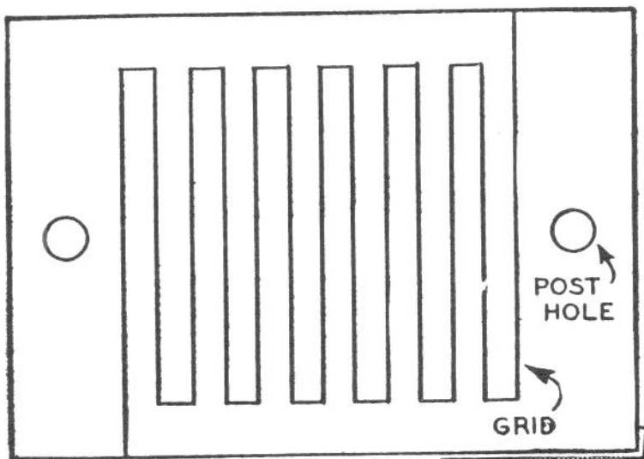


Above: Ruled lines are cut in graphite surface with sharp razor blade, using straight edge as a guide. From 16 to 20 lines per inch may be cut. Left: Lines scratched in graphite divide top of slate into 2 insulated parts.

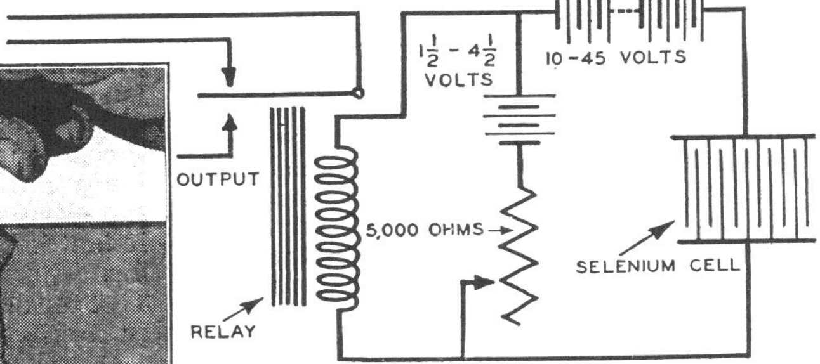


Diagram of circuit designed for use with this selenium cell. Relay should be sensitive enough to "kick over" at 1 or $1\frac{1}{2}$ ma. Dry cells operate relay, M battery the selenium cell.

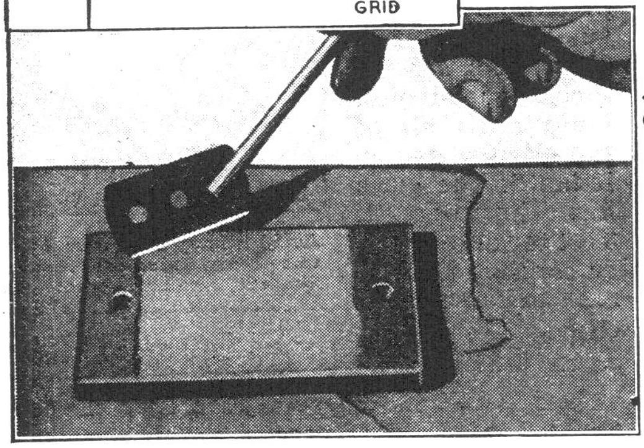

Slate covered with selenium is heated until chemical forms globules. Selenium is then rubbed back and forth across cell evenly with razor blade until selenium becomes gray.

leak is indicated, it may be eliminated by putting the cell across 110 volt line to arc out any offending particles; the grid is brushed gently with a camel's hair brush during the process.

High resistance leaks may be eliminated by putting the cell across the high tension terminals of a Ford coil.

### The Coating Process

The coating process may be carried out over a bunsen burner or gas stove. As the cells must be heated and cooled slowly, some material of sufficient thermal capacity to absorb sudden changes in temperature must be placed between the cell and the flame. A ¾ inch slate slab is good for this purpose, though a piece of fire brick flue lining, or even pie tins, one inside the other, with a layer of sand between, may be used.

A small quantity of selenium is placed on the gridded blank, and heat slowly applied, until the selenium just melts and tends to form globules. The selenium is then spread thinly over the gridded portion of the blank with a flexible razor blade mounted in a wooden handle.

The ends near the binding post holes should not be covered. The gas is turned down, and as the cell slowly cools the surface must be smoothed continuously and uniformly with the razor blade until the material thickens and the grey form of the selenium begins to appear. The surface, after turning grey, must not be touched. The layer of selenium must be kept as thin as possible, because the light penetration of this material is very small.

When the cell has greyed over completely, it is removed from the burner and placed in a hot oven, whose temperature is slightly below the melting point of selenium, 210 to

*(Continued on page 112)*

# Replace Burned Out Rectifier With Two Power Tubes

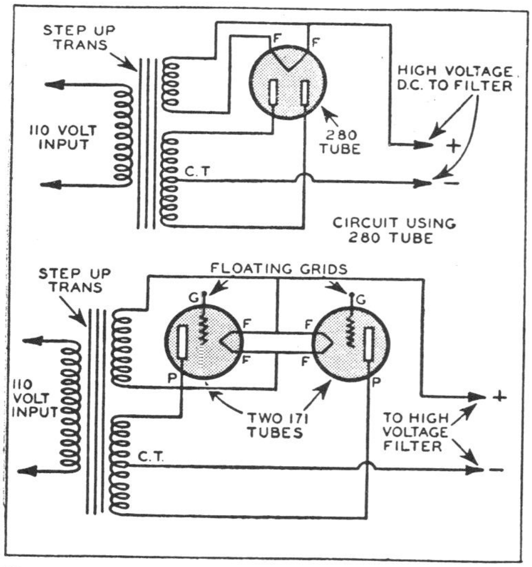


Filaments of 271 tubes are connected as illustrated above, leads being run to fil. transformer. Outer terminals of transformer then connect to plate terminals of each transformer.

HERE is a stunt that may help the radio fan who has burned out his 280 rectifier tube and has to send away for a new one.

First, disconnect the leads to the 280 socket. Then near by, on a separate piece of board if necessary, arrange two UX or UV sockets and connect their filaments in parallel.

Then connect one of the 280 filament wires to the plus side of the two sockets and the other filament wire from the pack to the minus side. Now disconnect the power wires connecting the plates of the 280 and connect one to the plate terminal of one of the new sockets and the other to the plate terminal of the remaining socket.

By using two 112 or 171 type tubes in the sockets you will get full wave rectification with sufficient current to operate a three or four tube receiver nicely.

## Extension Cord Holder Has Switch

A TOGGLE switch and double receptacle are an integral part of this extension cord holder. Designed to hold 50 feet of twisted 2 wire extension cord, it will be found useful in many ways about the home. Cord is unreeled as needed, and the bayonet plug then inserted in a convenient house receptacle. The toggle switch controls all appliances which are plugged into the holder.

This extension will be found especially useful for operating radios or speakers at various points inside or outside the home.

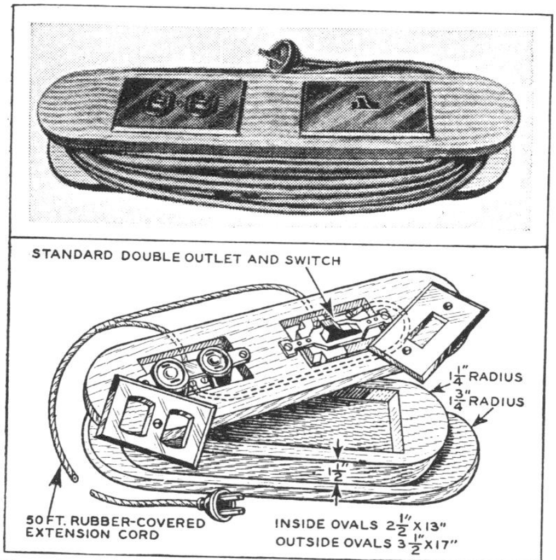


Completed holder takes 50 feet of twisted extension cord. Sketch below shows connections of receptacles to switch.

## Heat Glass to Pep Up Radio Tubes

THE new radio tubes with oxide coated filaments can be rejuvenated by slowly heating the glass with a blow torch or alcohol lamp until all of the silvery deposit inside the glass has disappeared. Test for shorts after cooling.

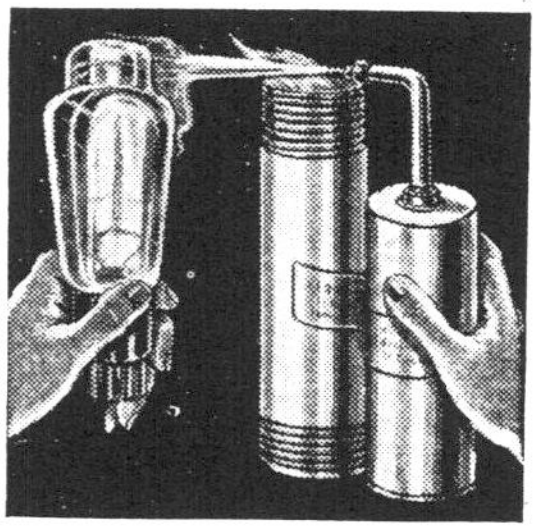

Avoid heating glass excessively when rejuvenating oxide-coated filament tubes by external heat.

## "Electric Eye" Opens Garage Doors

*(Continued from page 111)*

216 deg. Centigrade (410 to 420 deg. Fahr.). The temperature should be carefully maintained within these limits for two hours or longer, and the cell then allowed to cool slowly in the oven.

If the cell is properly made, it should show a dark resistance of around 20,000 ohms, and a conductivity in direct sunlight of 14 milliamperes at 45 volts.

If the cell shows the proper characteristics, it may be mounted in final form. A thin glass cover from a lantern slide cover glass or a photographic plate, should be bound over the selenium surface with adhesive tape. Binding posts should be fitted in the holes, using sheet copper washers to insure a good contact with the graphite.

These cells will, without amplification, operate the primary relays selling for a few dollars. Most of these "kick over" on 1 or 1½ M.A.

# Combination Switch Serves Same Purpose as Lock on Safe

A COMBINATION electric lock, which abolishes the need for keys, has proved itself very satisfactory.

An ordinary electric door opener is used, but this cannot be operated unless the person desiring admittance to the building or room presses the proper switches on an eight or ten gang push button panel.

The push buttons are arranged so that, when they are pressed, three will close the circuit and the other five will open it. If any buttons are pressed other than the correct three, the lock will not work. A switch may be installed inside the house or room, so that the entire electric lock can be disconnected at any time.

The combination electric lock is easily installed. The parts necessary are one ordinary electric door opener, one bell ringing transformer and an eight or ten gang push button. These should be wired as shown in the diagram, the power being taken directly off the 110 volt line. This power is stepped down in the belt transformer.

George H. Hill, of Williamansett, Mass., who has had such a lock on his workshop door for the past four years, says that numerous attempts have been made to enter, but that none have yet been able to find the correct combination of buttons which would open the door.

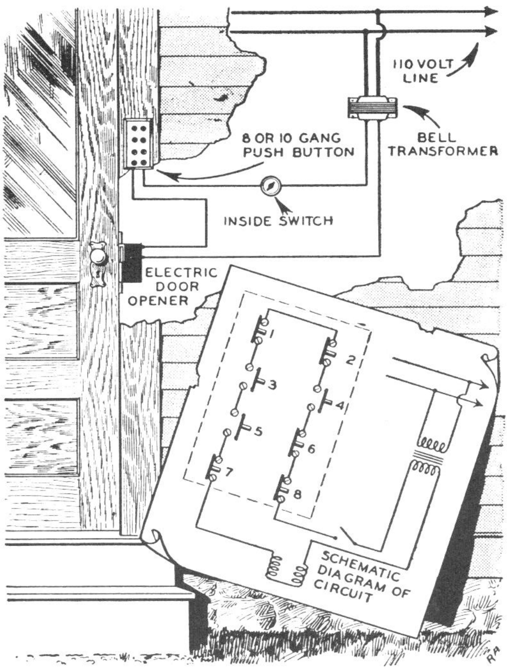


## Insulators at Center of Wire Make One Antenna Do the Work of Two

HERE is an old stunt for use where two neighbors wish to erect receiving aerials but the space between the two houses is limited.

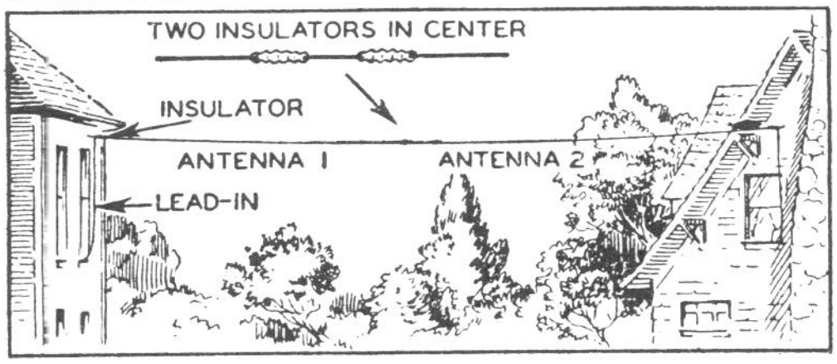


The space is measured between the spots for anchoring the lead-in ends of the aerials on each house. Then each aerial is cut slightly less in length than half this distance. Connect two insulators in series and fasten the end of an aerial to each end insulator. Then fasten an insulator to the other ends of the aerials and tie these in place on each house. Attach the lead-ins at these ends and connect to the receivers. With a spiral spring at one end the double aerial can be stretched tight and thus both parties can secure perfect reception.

## You Can Make a Ninety-volt B Battery Eliminator From Scrap Parts

FOUR chemical rectifier cells are used here to obtain full wave rectification of alternating current for a 90 volt radio plate supply. Make each jar exactly as specified for the battery charger on page 75; smaller aluminum plates may be used here, since the output current is smaller than in battery charging.

Be sure that the cells are arranged as indicated. Connect a small lamp in series with the line when first testing, to limit the initial current surge. After a few minutes the cells will be formed, and the lamp resistance may be reduced or cut out entirely.

Aluminum spoons or strips from aluminum pans can be used for electrodes. These will gradually wear away, but can easily be replaced.

A regular condenser and choke filter unit is needed here, just as with any other type of rectifier.

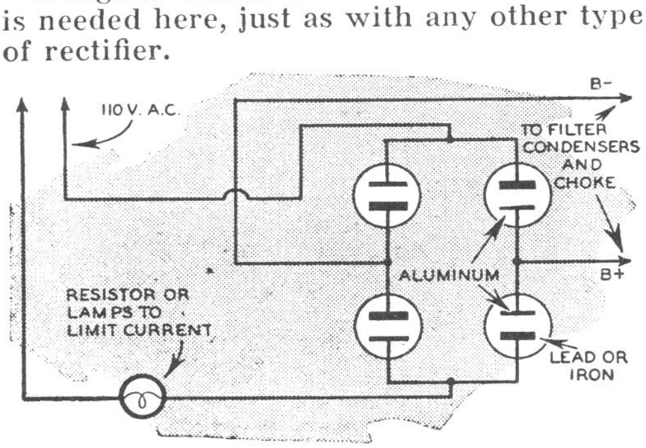

# Hear POLICE CALLS, Music, at Same Time

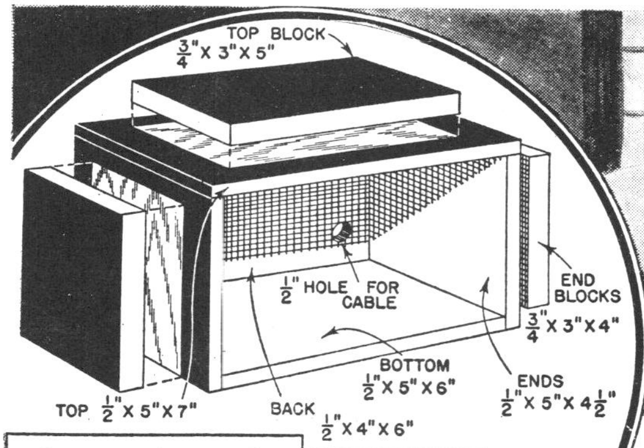


**Suggested modernistic cabinet for police call alarm is given above. Paint edges aluminum, sides glossy black enamel.**

**Mounted in attractive cabinet atop regular receiver, this police call alarm automatically brings in local police calls over regular programs. One tube circuit is used.**

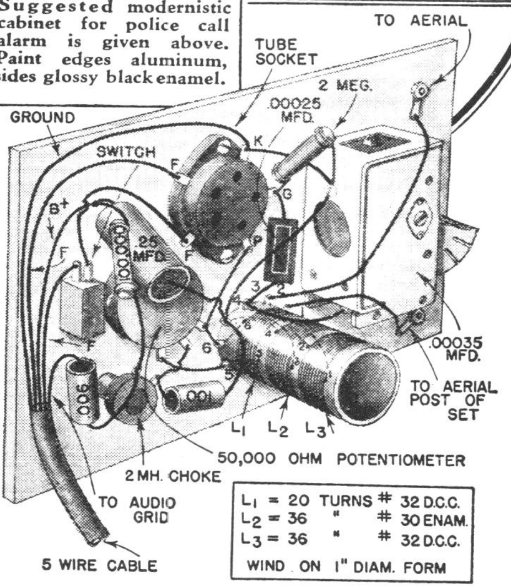


**Suggested panel mounting for parts is pictured above. Coil form can be 2½" long. Make 2 mh. choke coil by winding 200 turns of No. 32 d.c.c. wire, jumbled, on lead pencil core.**

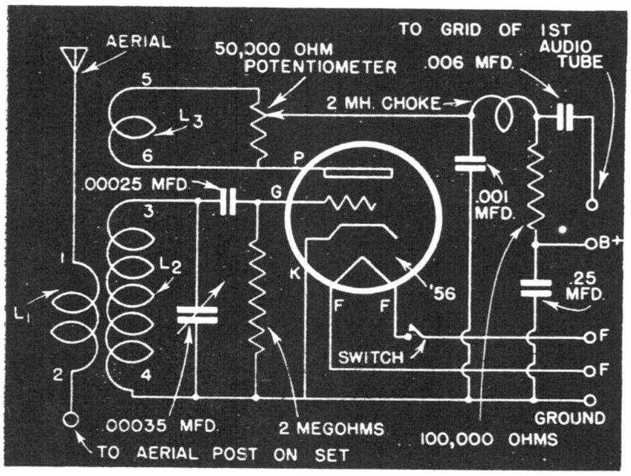


**Potentiometer may be any value from 50,000 to 250,000 ohms. Rating for .25 mfd. condenser is 400 volts. Other condenser sizes not critical—use closest size you have at hand.**

WITH this police call alarm connected to your radio receiver you can listen in to broadcast programs at any time, yet hear every police call sent out from your local police station without touching the receiver controls. These calls are heard with enough volume to make them easily understood regardless of the type of program being received, provided the police radio station is within ten miles of your home.

The alarm is simply a regenerative detector circuit tuned to the police wave band and coupled into the audio circuit of your receiver. All power for the alarm is taken from your set.

For sets with 2½ volt filament tubes, such as the '27, '24A, '35, '57, '2A5, etc., a '56 tube should be used. Tap off this filament power as near to the power transformer as possible. For sets with 6.3 volt tubes, use a '37 tube in the alarm.

To operate, turn on the receiver and the police call alarm. With no broadcast program tuned in, advance the potentiometer on the alarm until the hissing sound of regeneration is heard. Slowly tune the .00035 mfd. condenser until you hear a call from your local police station. Readjust the potentiometer for clearest reception of this police signal. No further adjustments are necessary. The filament switch of the alarm will turn it off at any time when police calls are not wanted.

The coil L1 is connected in series with the aerial of your receiver. Other connections to the receiver are made with a five-wire cable. B plus voltage is taken from the plate supply of your receiver by connecting to the plate of the output tube. A radio tube chart or manual will be of great help to you in making the necessary tube socket connections here.

# SHOCK TYPE Screwdriver Loosens Tightest Screws Easily

STUBBORN screws can be easily loosened with a new kind of screwdriver, shown in the accompanying photo. This tool has a blade at each end to take large and medium size screws.

Slipped over the screwdriver proper is a piece of iron pipe, both ends of which are cut half away so that the pipe can only turn on the screwdriver about half way before it is stopped by the pins.

When loosening a screw, the tool is gripped by the pipe and turned quickly. The pipe stops on the pins and delivers a sharp blow on the screwdriver, with a much better loosening effect than the steady twist of a regular screwdriver.

The screwdriver point may be made to fit loosely in the pipe to permit change of size. If you need a longer tool simply lengthen the pipe.

The pins are made of small nails inserted in holes bored in the points.

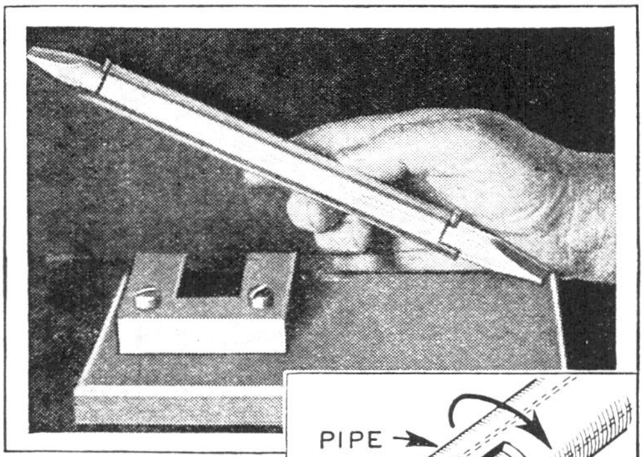

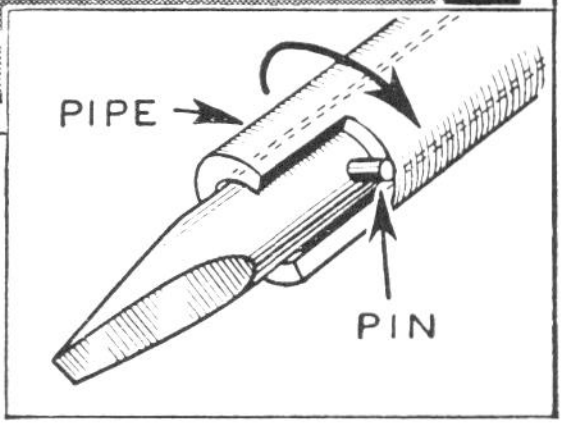


When screw is to be turned with this driver, the point is set in place and the handle given a sharp turn, communicating a telling twist to the screw. Right—How pipe handle is cut away.

## Caps Make Improvised Radio Set Dials

THE hard rubber caps found on tubes of tooth paste may easily be converted into small radio knobs such as are used for switches and volume controls. File rough notches in the shaft on which the knob is to be fitted. Fill the tooth paste cap with a good grade of cement, wood filler, or glue, and press it firmly over the shaft. Wipe off any excess cement and allow to dry. The many-colored hard rubber caps now being used on perfume and hair tonic bottles may also be used in this manner where somewhat larger dials are needed.

## Efficient Zepp Antenna Spreaders

MANY transmitting amateurs have difficulty in making spreaders for the two feeder wires. Here, however, is one way to make such spreaders which will bring returns to the builder in increased efficiency.

Cut as many pieces of hardwood as needed, 12 in. long by ½ inch thick (squared stock). Saw a slot in each end of the spreaders 1 inch long, ending in a hole the same diameter as the feeder wires. Then drill a hole through the ends of the spreaders as illustrated in sketch. For insulation boil spreaders in paraffin.

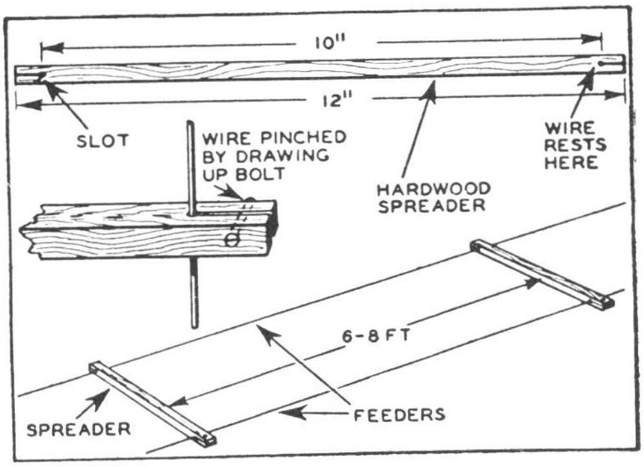


## "Pigtail" Leads Stop Tuning Condenser Noises

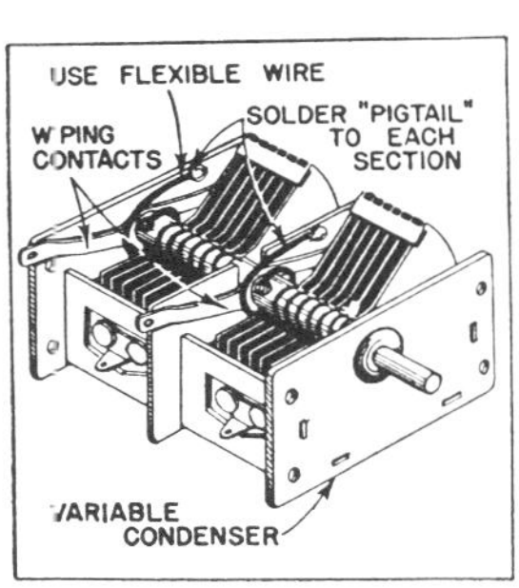


THE wiping contacts connecting the rotors of variable tuning condensers to the frame often become dirty, producing harsh noises and sometimes oscillation when tuning from station to station. One way to cure this trouble is to solder flexible wire or "pigtail" leads from the condenser frame to the rotor, bridging the wiping contact. In gang tuning condensers it is important to do this to each section. Leave the old contacts in place, as they provide proper tension and keep the plates properly spaced. To test for rubbing plates, connect headphones in series with a dry cell between the stator and rotor. Clicks in the phones indicate the plates are touching.

# Car Dash Panel Is Baffle Board for Radio LOUD SPEAKER

Photos show front and back of auto radio speaker mounted to use dash panel as baffle board. Grill goes on inside of car, over hole cut in dash.

MOUNTING the speaker of a car radio on the engine side of the dash panel allows more room inside the car, and at the same time provides a large baffle board for the speaker itself, greatly improving tone quality.

First cut a hole in the dash about 1" less in diameter than the speaker. Remove the screen or grill, then bolt the speaker solidly in place on the engine side. Replace the speaker grill on the other side of the dash panel, either by gluing or bolting it in position.

In cutting the large hole in the hard steel panel, drill a series of holes as close together as possible around the circumference, using a power drill, then cut out the disc with a cold chisel.

Low note response of the speaker will be greatly improved by this improvised baffle board arrangement.

## Neon Lamp Acts as Station Finder

THIS neon lamp tuning indicator will prove a valuable aid in locating foreign stations on all-wave radio receivers not having a beat oscillator. The indicator produces a faint hum in the loudspeaker continually; when a station is found the hum will become much louder. The hum can be eliminated by disconnecting the neon lamp, once the station has been accurately tuned in.

The small neon glow lamp is connected across cathode and plate of either the first R. F. or the first I. F. tube in the receiver, as indicated in the sketches.

By changing the size of the fixed condenser which is connected across the neon lamp, the pitch of the hum can be changed.

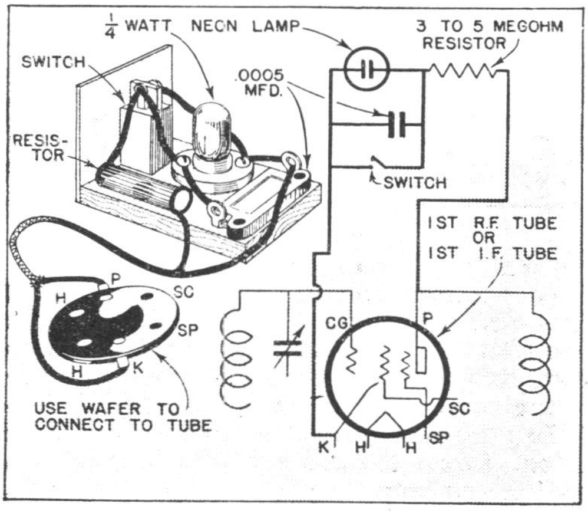


Circuit diagram and suggested assembly of neon lamp tuning indicator. Use wafer made from fiber discs to connect to set, after getting proper tube connections from tube manual.

## Shield Leads With Wire Springs

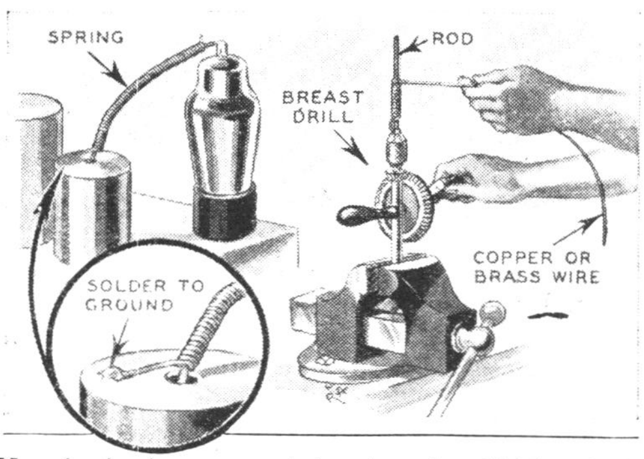


Here is simple way to wind springs for shielding leads. Ground both ends of spring by soldering to chassis or slipping under convenient bolt on chassis of radio set.

IN SHIELDING wire leads for radio work, coil springs offer many advantages over common metal braiding. Since they can easily be wound to any desired length and diameter, capacity effects can be eliminated where necessary. The springs are easy to slip over soft wires, and will support the leads.

The spring is made by coiling brass or copper wire on a rod of the correct diameter.

Either copper, iron, or spring steel wire may be used for shielding, though copper will undoubtedly be easiest to work with. Remember that the springs will expand a little after winding.

# Tested Kinks Speed Experimental Radio Work

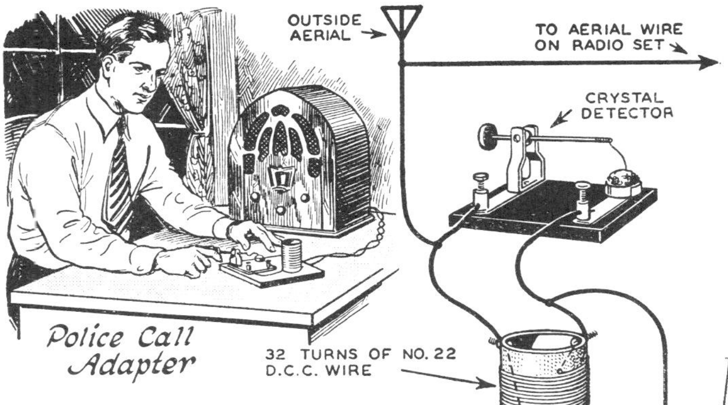


Novel coil-crystal adapter brings in police calls on many types of radio sets. Connect coil in parallel with any type crystal detector, across aerial and ground of set. Tune radio receiver as usual. When police call is received, adjust catwhisker on crystal until signal is loudest and clearest.

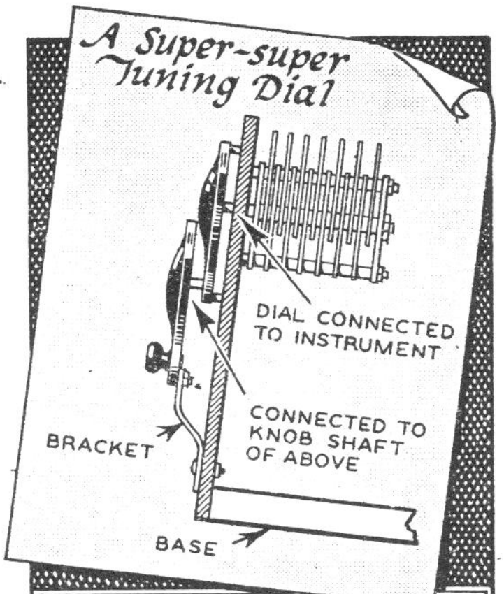


Even the ordinary vernier dial is often not fine enough for tuning on certain short wave bands. This double vernier dial will do the trick. Mount tuning condenser higher than usual. Attach dial but remove vernier knob. Attach second dial to vernier shaft, mounting dial on stiff metal bracket. Crowded wave bands will be spread out greatly by the second dial.

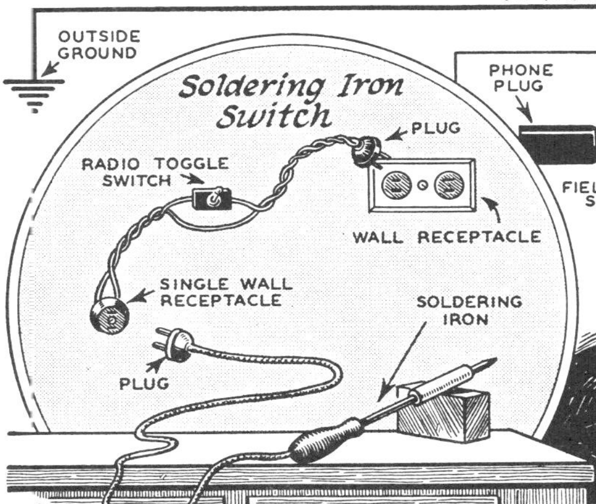


Cord and switch keeps soldering iron ready for instant use. Mount single screw wall receptacle and radio toggle switch on bench where soldering is done. Leave both plugs in receptacles at all times, so switch controls iron.

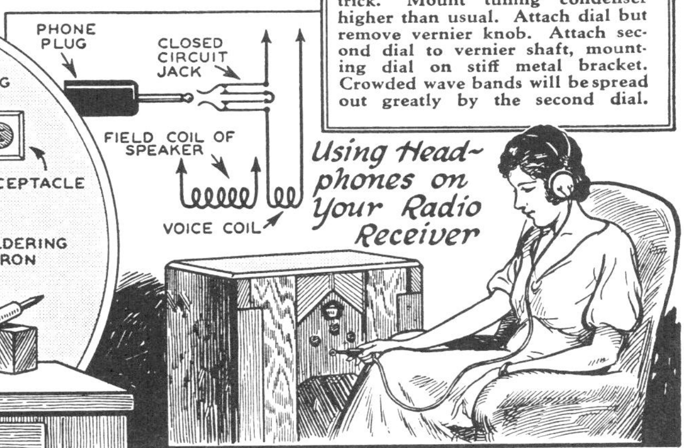


Connect phone jack in series with dynamic speaker voice coil. Since coil resistance is 15 ohms, phones 2000 ohms, speaker will not work when headphones are used on radio.

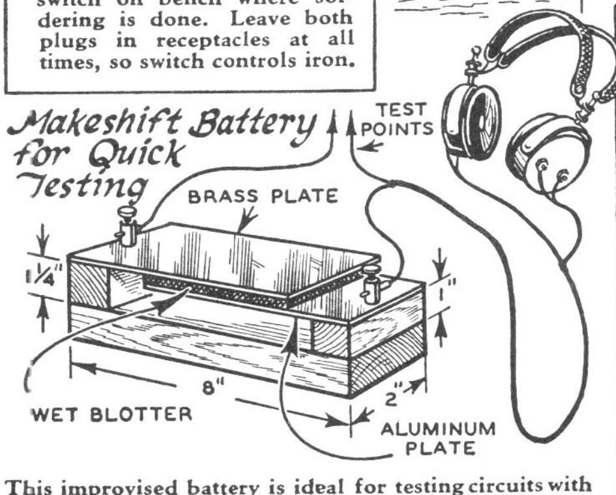


This improvised battery is ideal for testing circuits with headphones. Mount sheets of aluminum and brass as shown, pressing them together so that plain blotter just fits between. With blotter saturated with salt solution, enough current flows to make click in phones.

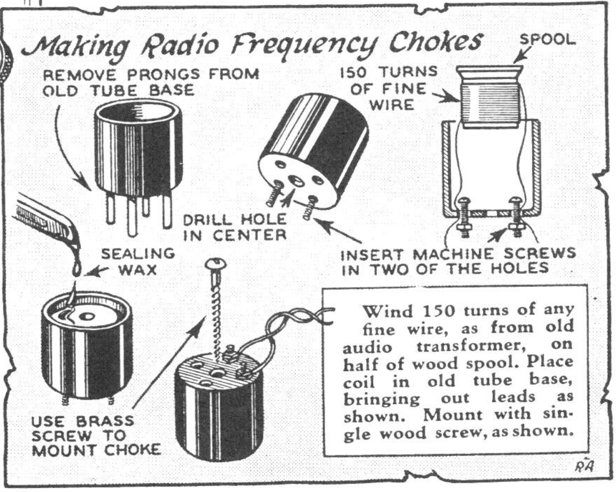


Wind 150 turns of any fine wire, as from old audio transformer, on half of wood spool. Place coil in old tube base, bringing out leads as shown. Mount with single wood screw, as shown.

# Two-Tube All-Wave Radio

**By using the new triple-function 6C6 tube, foreign reception with headphones, local programs on the loudspeaker are obtained with this compact two-tube all-electric long and short wave receiver.**

IN ADDITION to bringing in broadcast entertainment, this compact two tube all-electric receiver provides the thrills of foreign short wave reception. It will work anywhere where 100 to 120 volt power, either alternating or direct current, is available.

The set is built up on a metal chassis cut from aluminum or any available sheet metal. The metal front panel also serves as the front of the ornamental crackle finish metal cabinet used so extensively by short wave set builders. The entire set, with tubes, coils, and headphones, weighs less than four pounds.

Though designed for headphone operation, this set will operate a small magnetic speaker very nicely on the powerful local stations. Two or more sets of phones may be employed at the same time if desired. For foreign reception a good outdoor antenna with one of the commercial noise-eliminating lead-in systems is recommended, but for local broadcasts even a short indoor aerial will prove satisfactory.

The circuit is beautifully simple. Only two tubes are used, one of these, the type '37, with plate and grid tied together, serving only as a rectifier.

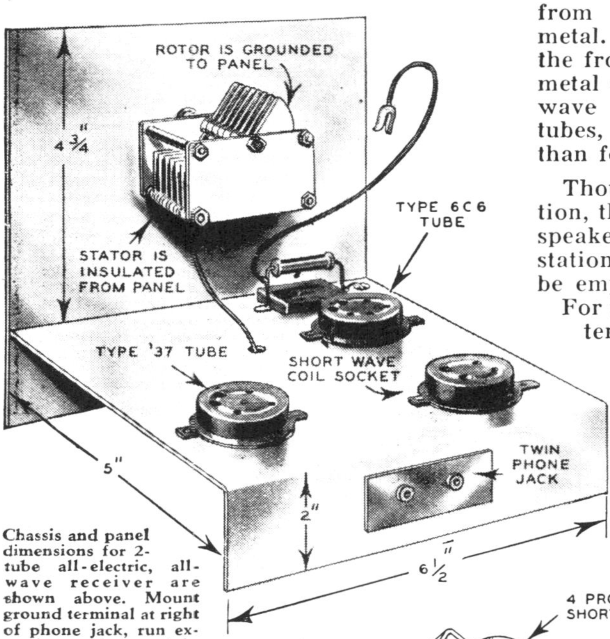


**Chassis and panel dimensions for 2-tube all-electric, all-wave receiver are shown above. Mount ground terminal at right of phone jack, run extension cord out at left.**

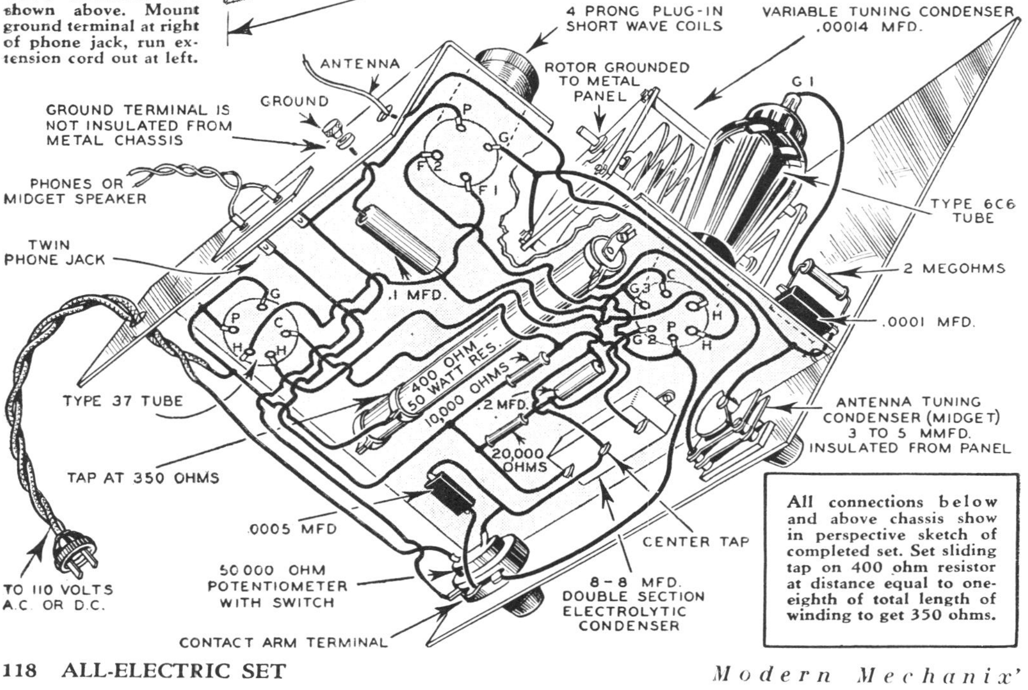


**All connections below and above chassis show in perspective sketch of completed set. Set sliding tap on 400 ohm resistor at distance equal to one-eighth of total length of winding to get 350 ohms.**

# Plugs Into LIGHT SOCKET

The new 6C6 tube used in this set actually does the work of three ordinary radio tubes, picking up the radio frequency signals, detecting them, then furnishing the audio frequencies for the headphone.

Naturally, regeneration is used to make the set as sensitive and selective as possible. A potentiometer shunted across the tickler winding of the plug-in coil serves as regeneration control.

The Cisin Universal A.C.—D.C. circuit permits this set to be used interchangeably on alternating or direct current. The filaments of the two tubes are connected in series with a 350 ohm voltage limiting resistor. In case a pilot light is desired for the tuning dial, this is connected in series with the tube filaments and shunted by a 30 ohm wire-wound resistor. The voltage limiting resistor should then be set at 320 ohms.

A small trimmer condenser in the antenna circuit permits use of either a long or short aerial, and gives an added tuning control on weak short wave stations.

*(Continued on page 120)*

Regeneration control at lower right and large vernier tuning dial are main controls on this short wave set.

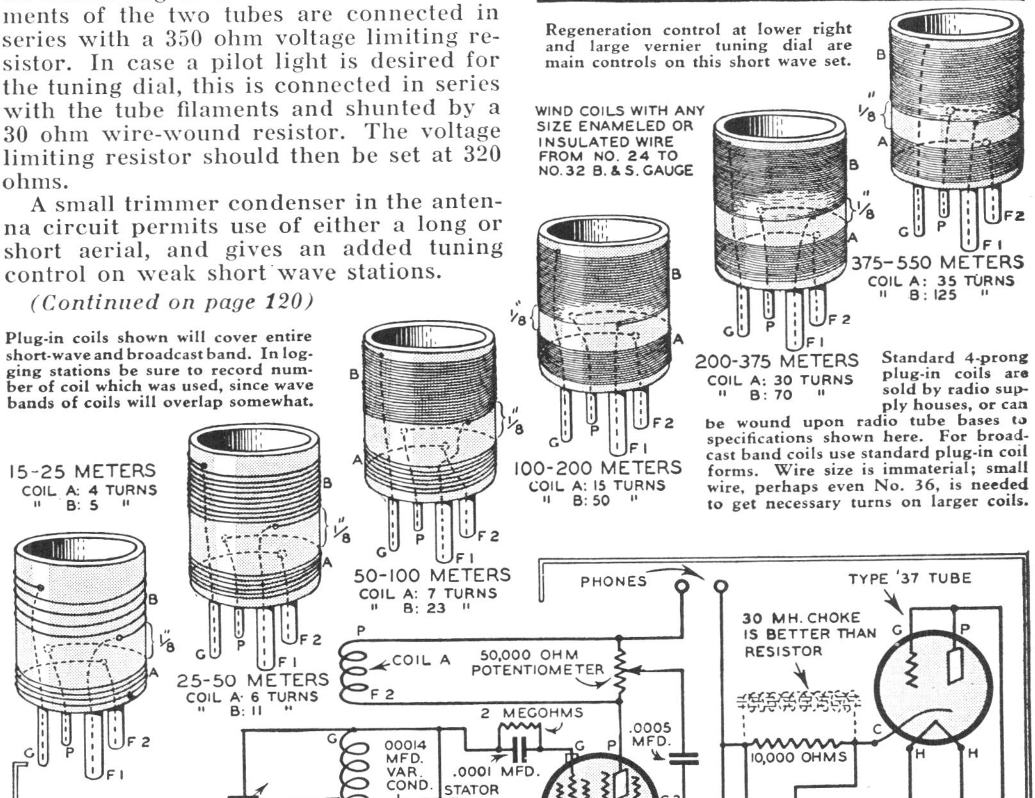


Plug-in coils shown will cover entire short-wave and broadcast band. In logging stations be sure to record number of coil which was used, since wave bands of coils will overlap somewhat.

Standard 4-prong plug-in coils are sold by radio supply houses, or can be wound upon radio tube bases to specifications shown here. For broadcast band coils use standard plug-in coil forms. Wire size is immaterial; small wire, perhaps even No. 36, is needed to get necessary turns on larger coils.

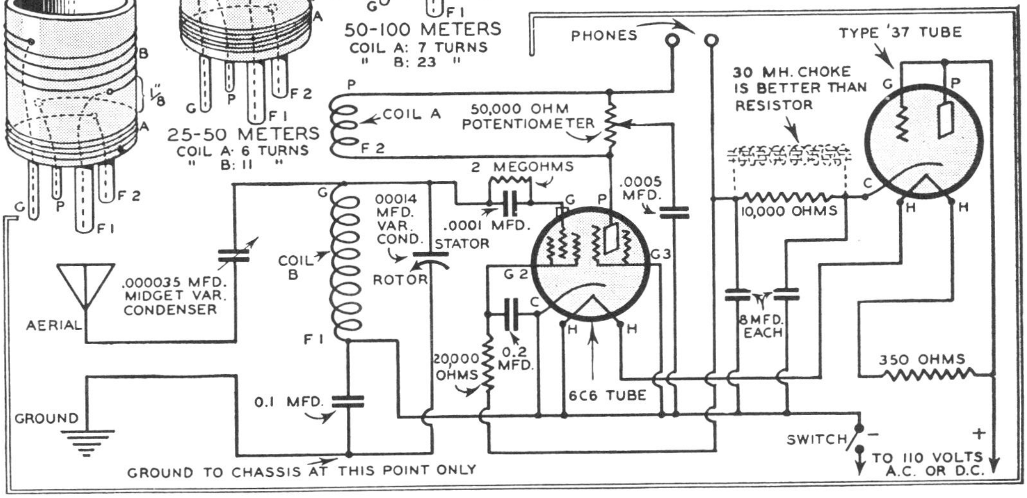


Schematic circuit diagram of set is shown here for those who prefer to plan their own arrangement of parts. Filter resistor of 10,000 ohms may be replaced by standard filter choke if a.c. hum comes through phones. In tuning set, locate station with tuning dial, then adjust regeneration potentiometer until set is just ready to break into regeneration squeal. Antenna condenser need be tuned only for weak short wave stations. Note that set wiring is grounded to chassis at one point only.

# Ideas for the Home Electrician

## *Charge Battery With Ford Generator*

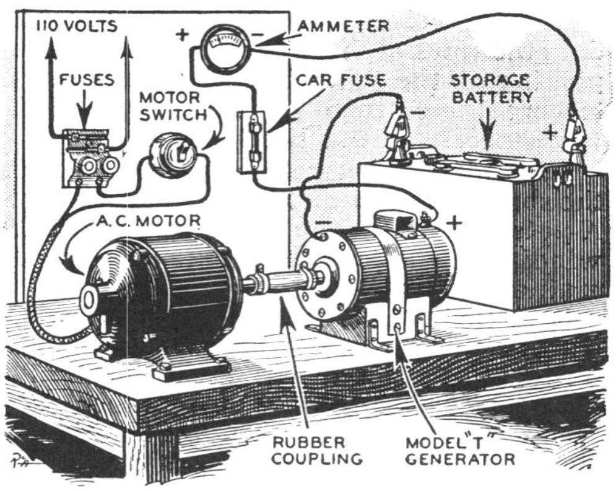


A MOTOR generator set for storage battery charging will be a valuable asset in any home workshop. Mount an old a.c. motor of about ¼ h.p. size on a bench, in line with a Model T Ford generator, and couple the two shafts with a short length of rubber hose. Leave the cutout on the generator. On a small panel in back, mount fuse blocks for motor and generator, a switch for the a.c. motor, and an ammeter to indicate the charging rate of the generator. The metal frame of the generator is negative so connect this to the minus post of the battery. The terminal on the cut-out is connected through ammeter and car fuse to the plus post of the battery. Batteries can be completely charged in 6 to 8 hours.—J. R. McDermott.

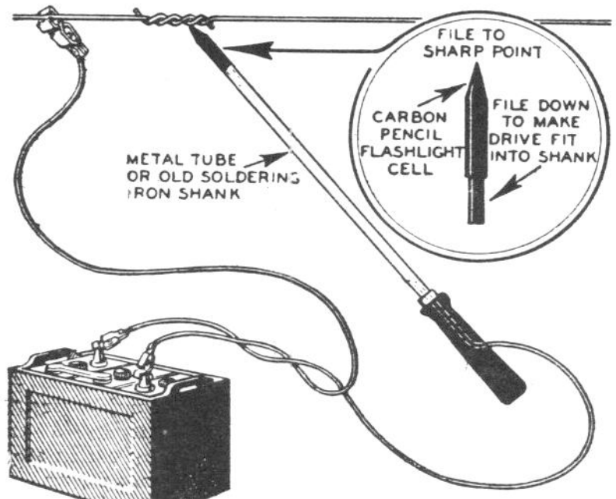


## *Making a 6-volt D.C. Soldering Iron*

A SIMPLE soldering iron requiring only a six volt storage battery is easily made. Remove the center carbon from a flashlight battery, leaving the metal cap in place. Grind or file to a point the lower end of the carbon and solder a length of insulated wire to the brass cap. For convenience carbon may be mounted on old electric soldering iron. Connect the carbon electrode to one post of battery and the piece being soldered to other, using battery clips. Clean joint, apply flux, bring carbon point in contact with work, draw back a little, and apply solder as soon as arc has heated work. —L. B. R. Jones.

## *"Feet" for Radios Made from Rubber Heels*

IN BUILDING radio and electrical apparatus, it is advisable to use rubber cushions to prevent cabinets from scarring table tops. Rubber heels cut apart are ideal for this purpose. With a razor blade cut the heels apart so that each piece contains one hole. Trim the pieces to a cylindrical and slightly conical form, and smooth with sandpaper. Attach the feet with round-headed screws, drawing the head well up.—Leroy Vigna.

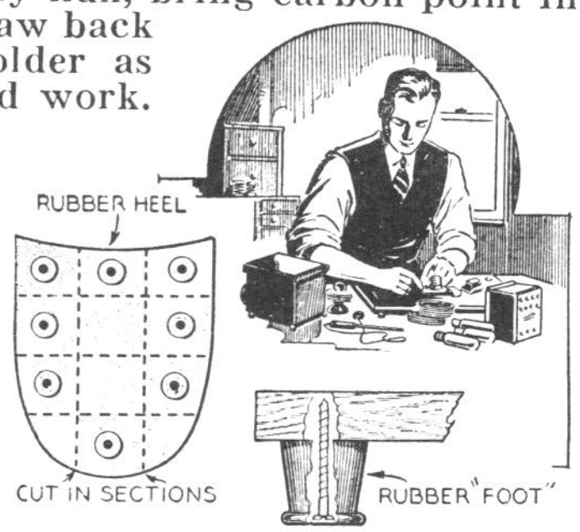


## Two-Tube All-Wave Receiver Plugs Into Light Socket

*(Continued from page 119)*

A .1 mfd. condenser isolates the ground connection and chassis from the power line, preventing short circuits which might burn out the house fuses.

While a small 10,000 ohm metallized resistor, by-passed at each end by an 8 mfd. electrolytic condenser, is used here as a filter pack, some may prefer to use an old audio transformer or standard 30 millihenry choke in place of the resistor.

After fastening the chassis to the front panel with angle brackets, mark the location of each part and drill the necessary holes. Mount the three tube bases, antenna and tuning condensers, potentiometer and phone jack, then begin with the wiring.

A shield should be used over the 6C6 tube for noise-free reception. This can be bought at a radio supply house for about ten cents.

Complete coil winding data has been given for those who prefer to wind their own plug-in coils. Tube base forms can be used for the short wave coils, but these will be far too small for the number of turns needed on broadcast band coils. Shellac all coils after winding to keep the turns in position.

# INVENTORS

## *act now to* PROTECT YOUR IDEAS WITH A U.S. PATENT

**MAIL COUPON FOR MY FREE BIG 48 PAGE BOOK AND FREE FORM FOR RECORDING YOUR INVENTION**

CLARENCE A. O'BRIEN
Registered Patent Attorney

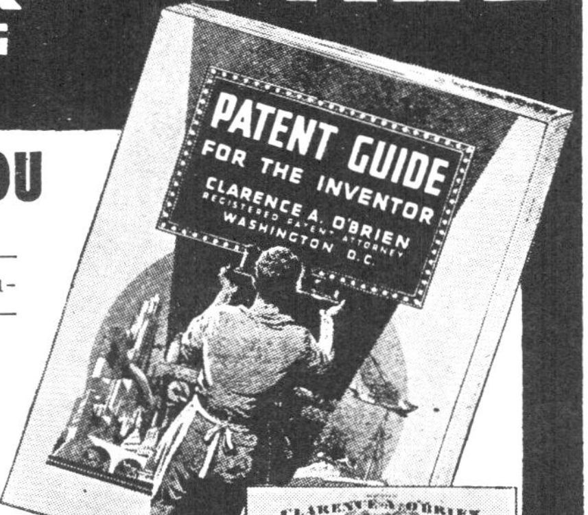


### MY BIG BOOK TELLS YOU

**The Importance of Quick Action**

If you have a new idea that is patentable—and want to secure for yourself your inventor's sole right to profit from the idea—you can't afford to lose time in securing the protection of a U. S. Patent. Other inventors may be working on the same idea right now. Should they develop this idea and file patent application before you do—the burden of establishing priority will be upon you. My Big Book—"Patent Guide for the Inventor"—will tell you the quickest possible way to secure Patent Protection for your idea. SEND FOR IT TODAY!

**The First Important Step in Securing a Patent**—is to obtain the services of a Registered Patent Attorney. The U. S. Patent Office advises that you secure the aid of an experienced, **Registered** Patent Attorney to prepare your statements. Establish the date of your invention as explained in My Book and disclose your idea to him in correct form. He then makes a SEARCH of the Patent Office Records and reports to you on existing U. S. Patents which might bear on your invention, giving his opinion on the advisability of proceeding and filing formal application.

SMALL IDEAS MAY HAVE LARGE COMMERCIAL POSSIBILITIES. Ideas do not have to be complicated to be profitable when patented. The 5c and 10c stores flourish on simple patented products which are commercially successful. The zipper, safety pin, thumb tack, paper clip, collar button—all are simple yet familiar inventions—brought their inventors rich rewards.

**FREE Record of Invention Form**

Use my FREE "Record of Invention Form" to assist you in recording the descriptive outline of your idea and to establish the date of your invention. With it you can make a rough sketch of your idea and have it witnessed in a manner of value to you as evidence should occasion arise.

**Reasonable Fees—Convenient Terms**

Don't worry about finances. My fees are very reasonable—I can arrange convenient terms of payment. It is my custom to permit clients to pay for my services in installments as the application progresses in my office. See My Big Book for fees in obtaining Patent. It explains my arrangements clearly. **Send for it NOW.**

**Request My Book—NOW!**

Know exactly what you should do to secure a Patent—what must be done every step of the way—how the Patent Law protects you—what I, as a Registered Patent Attorney, can do for you. **Mail the coupon NOW.**

**CLARENCE A. O'BRIEN**
*REGISTERED PATENT ATTORNEY*
635-A Adams Building
WASHINGTON, D.C.

Please send me at once your Big Free 48-page Book, "Patent Guide for the Inventor," and your Specially Prepared "Record of Invention" form. This request does not obligate me.

Name ..............................

Address ..............................

..............................

(Please write or print plainly)

THE
HARRISON
RADIO
COMPANY

CARRIES ONE OF THE LARGEST AND MOST COMPLETE STOCKS OF RECEIVING AND TRANSMITTING APPARATUS IN THE WORLD! SEND IN YOUR LIST OF PARTS FOR OUR MONEY SAVING QUOTATION BY RETURN MAIL!!

Our years of experience enable us to intelligently and promptly fill your every radio need. Try us —we know you will be satisfied!

WE DISTRIBUTE EVERY NATIONALLY KNOWN LINE OF SETS AND PARTS INCLUDING—
HAMMARLUND—NATIONAL—ROYAL
RCA — CARDWELL — SYLVANIA
BURGESS — CENTRALAB
CLAROSTAT — FULTONE
JOHNSON — RCA
DeFOREST, Etc.
LOWEST PRICES FASTEST SERVICE

SHORT WAVE RECEIVERS—
TRANSMITTERS—AMPLIFIERS
Priced $3.00 to $300.00

Our Easy-Construction Kits take the guess work out of building your own equipment! Kits are complete down to the last screw —Professional appearing metal chassis and panels with all holes—Factory-wound coils —Clear, simple instructions with pictorial and schematic diagrams and numerical "E-Z Wire" system! Inexpensive wiring and testing service.

SEND FOR FREE CATALOG

HARRISON RADIO CO.
142 Liberty St., Dept. MXM, New York City
VISIT OUR NEW LARGE SALESROOMS

WORLD WIDE ALL WAVE~ RADIO $1.95

Get stations all over the world with this POWERFUL radio receiver! All Wave—15 to 600 meters! Works on two inexpensive batteries! NOT a toy or attachment! Results Guaranteed! Ace kits include all parts mounted on attractive metal chassis and panel. Clear picture instructions. Wire it yourself! It's easy!
1-TUBE KIT—$1.95 * 2-TUBE KIT—$2.85
ALL ELECTRIC 105-125 volt AC-DC two tube house current set. No batteries needed! Complete Kit $3.65.
Kits wired 75c extra. Tubes 85c ea. Double Phones $1.25
ORDER NOW! Send $1. balance C.O.D., or postpaid if full remittance. Postcard brings free literature.
FREE!! Large three color Map of the WORLD, valuable Short Wave data, etc.!! Send 10c for handling, NOW!
ACE RADIO LABORATORIES
1619 Broadway Dept. RBM New York City

## Notched Dials Locate Stations

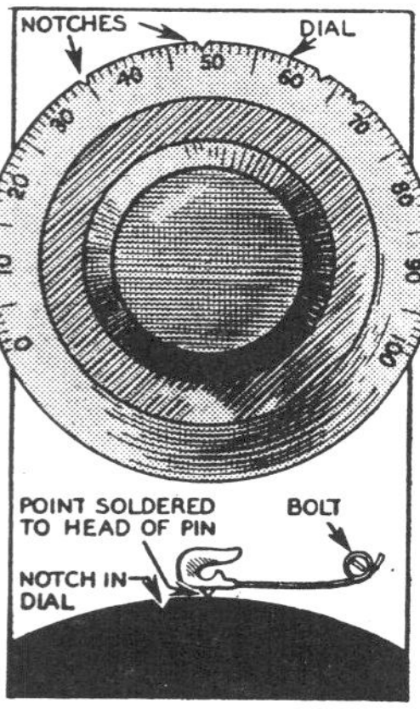


THERE are many times when the operator wishes he knew at just what setting of the dial he had received certain stations. This is a simple matter to keep fixed by making very fine notches in the dial edge and having them come in contact with a fixed point on the panel.

Make your notations on the dial at the exact point where the station comes in loudest. When these points have all been determined remove the dial and cut very small notches in the dial edge as shown.

Replace the dial and then mount a safety catch on the panel as shown. A bit of brass should be soldered to the under side of the pin and filed to a knife edge. This will spring down into the notch when the spring tension of the pin is sufficient.

File notches in the dial to locate immediately the station you want.

## Wind R. F. Chokes on Silk Spools

R. F. CHOKES made by winding the wire on one form are inefficient because of high capacity. To distribute the capacity wind the coil in sections as shown in the sketch below.

Get a number of small wooden silk spools, as many as are needed, and also a length of wooden dowel. Smear the dowel with glue and force on the spools as illustrated. Drill a small hole through the two end spool flanges for the wire to pass through and then put on the windings.

Having estimated the number of turns your particular choke calls for, divide this number by the number of spools you have in the form and wind on just so many turns to a spool. This distributes the total capacity and thus raises the efficiency of the choke.

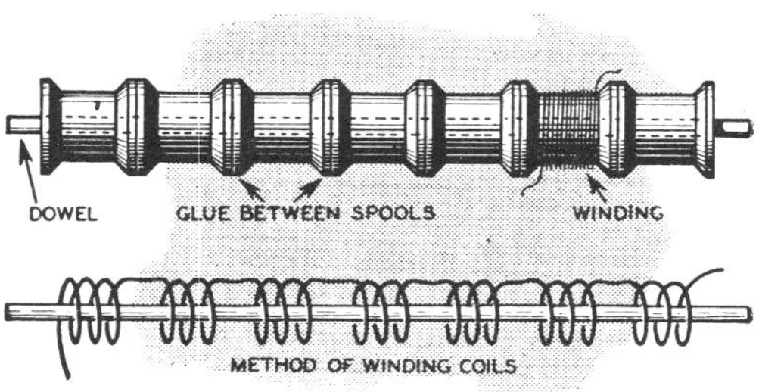

# TRUE STORIES by GRADS

# FOLLOW BILL MOORE'S EXAMPLE

MAIL TODAY

**You, too, can be another Bill Moore—in the BIG PAY class! Radio and Television offer a brilliant future—but the TRAINED TECHNICAL men get the BIG PAY jobs! First National Television makes you a radio and television expert in six months! No previous experience needed! Thorough training in our ultra-modern laboratories and actual work, for qualified students, in our 1000-watt commercial radio station W9XBY and television experiment station W9XAL give you experience which means preference in any radio job.**

**Tune in on Hi-Fidelity Station W9XBY, 1530 kilocycles**

**License While Training**—A government operator's license is a passport to a real job with a future! My students qualify for their licenses and get certified service records while training!

**The Way to BIG PAY!**

Clip this coupon and MAIL IT TODAY! It can be the biggest step in your life—it will lead the way to highest paid radio jobs.

S. Q. NOEL, Pres. First National Television, Inc. (Training Div.) Dept. KK-4
Power & Light Bldg., Kansas City, Mo.

Without obligation, send me postpaid FREE Illustrated Folder telling about new opportunities in radio and television. I am 17 years or older

Name__________ Age______

Address__________


## An Inexpensive, Low Current Power Pack for the Beginner's Transmitter

A TRANSMITTING amateur, just entering the game and finding himself short of power supplies, can use the power pack shown in the accompanying diagram with assured success. The writer used this same hookup to operate a pair of 245 tubes as oscillators, and on 40 and 80 meters, obtained reports of steady signals, pure DC and ranging from R5 to R8 in volume.

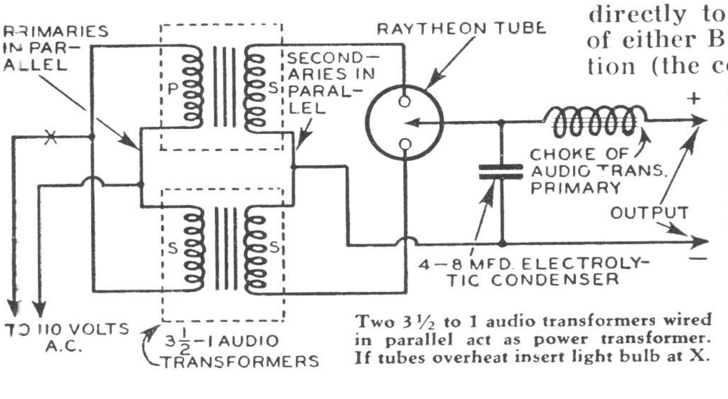


Two 3½ to 1 audio transformers wired in parallel act as power transformer. If tubes overheat insert light bulb at X.

The power transformers consist simply of two matched audio transformers.

The outside ends of the turns connect directly to the plates of a Raytheon tube of either B or BH type. The third connection (the cold anode) connects to one side of the filter choke and one side of a suitable fixed condenser (cost half dollar) as shown. The choke can be the primary of a heavily constructed audio transformer.

With 110 volts input, using 3½ to 1 transformers, you will get about 350 volts output.

## *Alarm Clock Switch Turns Off Radio Receiver*

WITH this easily made alarm clock radio switch, you can be lulled to sleep with strains of music from your broadcast receiver. To rig up the device, fasten the clock securely to the board with a loop of wire as illustrated in the accompanying drawing. Now a toggle switch should also be secured to the base board about three inches in back of the clock.

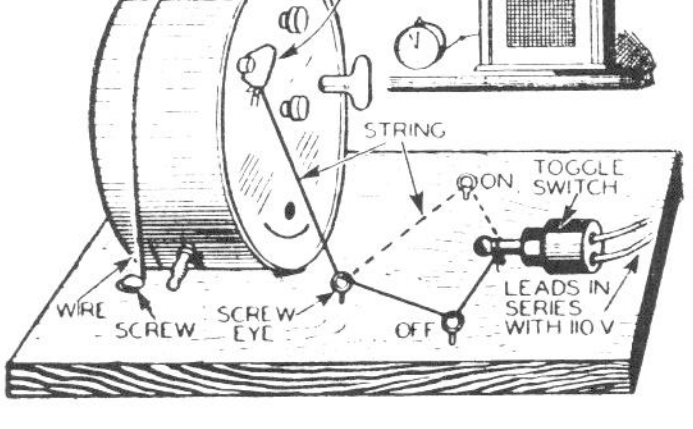

# Filter Out STATIC from RADIO Antenna

by THOMAS A. BLANCHARD

Installation of static-free antenna and line noise filter are shown in above sketches. Mount filter in neatly constructed metal box.

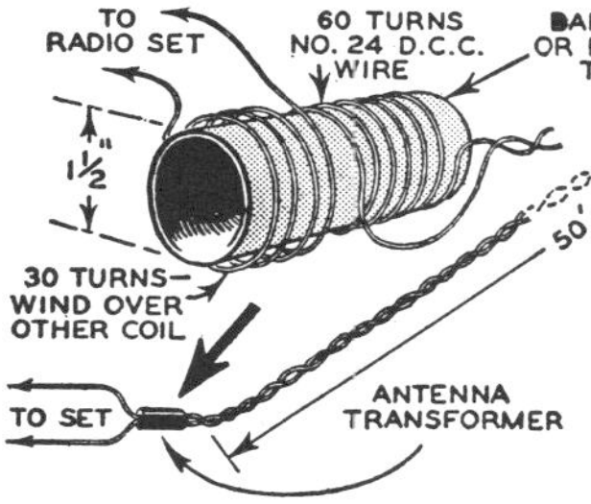


Twisted lamp cord dipped in hot paraffin serves as lead-in for special antenna. Transformer coil is placed at radio set.

Keep antenna high as possible, well-insulated at every point. To match antenna perfectly to receiver, wind antenna transformer coils side by side, add 10 turns to small coil, and use pair of sliding arm contacts to change turns.

MAN-MADE radio static, entering your radio receiver either through the antenna or by way of the power line, is not at all difficult to stop. A special antenna designed to pick up only signals within the frequency range of broadcast stations will cut out interference developed by motors, telephone dials, and other electrical devices.

A capacity-inductance filter in the power line running to the receiver will stop unwanted electrical disturbances from that source. While it is as yet impossible to stop all natural static, it is generally man-made static which gives the most trouble.

The line filter, consisting of two coils each in series with a power line lead, and two condensers in series across the line, with their midpoint grounded to the metal case and to a water pipe, is plugged into a base receptacle. The doublet antenna is not at all hard to erect.


Be a RADIO EXPERT - Learn AT HOME

Be the ONE Man in 1000 Who Can Service Modern Sets!

FREE! OF EXTRA COST
To start you making money without delay we equip you with this Circuit Analyzer and Point-to-Point Resistance Tester.

Modern circuits stumps the old-time cut-and-try radio fixer. More puzzling complications are coming. Service men with up-to-the-minute knowledge are urgently needed. Such men easily make up to $3 an hour.

No Past Experience Needed

Past experience counts for little at this time because changes in radio design are so swift and radical. That's why you, as an R. T. A. Certified Radio-Technician can step right into the big money at once. Write today for full particulars.

EARNS $2.00 to $5.00 AN HOUR
I started a radio shop handling all kinds of radios and electric equipment. Since starting this shop, I have been able to earn from $2 to $5 an hour. I have built up a good reputation from sales and service work. These satisfied customers have referred their friends to me and as a result I have all the business I can handle. R.T.A. sure opened my eyes to big money and I want to thank you for all you have done for me. My income is more than double what it was at the time I enrolled and I expect to earn still more.
*Ernest Lanctot*

RADIO TRAINING ASSOCIATION OF AMERICA
Dept. RBM-5 4513 Ravenswood Avenue Chicago, Ill.

## Condenser Reduces A. C. Voltage

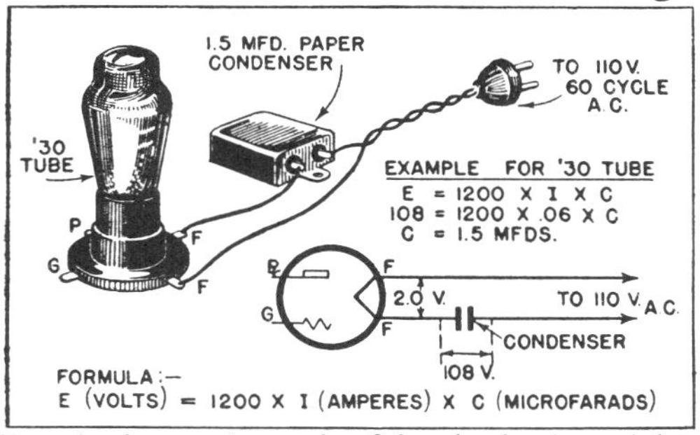


**Formula gives condenser size C in microfarads needed to pass given current I and drop voltage required amount E.**

IN THE same manner as resistors are used to reduce direct current voltages for radio tube filaments, filter condensers can be used to reduce A. C. line voltages to the correct values for radio tube filaments. A voltmeter should always be used to check the voltage at the filament. Smaller condensers can be connected in parallel to accurately adjust this voltage, since the exact condenser called for by the equation will seldom be available.

## Hitch-Hiker's Portable Radio

*(Continued from page 13)*

vide a sound outlet for the 5" magnetic speaker. Drill a small hole at the rim to bolt the speaker rigidly behind the baffle plate. The opening may be covered with a grill cloth to protect the fragile speaker cone from injury.

Other coils, with fewer turns on both primary and tickler windings, may be wound to cover the different short wave bands. The number of turns is not very critical—a little experimentation will enable you to wind a coil for any desired short-wave band.

The midget speaker has a cone diameter of five inches, and will just fit into the shield can. For best results, it should be of the permanent magnet type, with a D.C. resistance of about 1000 ohms, and an impedance of 7000 ohms to match the recommended output load of the '33 power tube.

The potentiometer is a midget of 50,000 ohms maximum resistance. This instrument acts as a combined regeneration and volume control.

If desired, a 6-ohm variable resistor may be used in the positive filament lead of the '33 tube to give a better volume control; this would necessarily entail further drilling of the panel and another control knob.

**BLUEPRINTS for the Hitch Hiker's two-tube portable receiver, containing greatly enlarged diagrams, are available at 50c postpaid. Modern Mechanix Publishing Co., 529 S. 7th St., Minneapolis, Minn.**


*Now—a high-Powered—*

# Radio Engineering Library

***—especially selected by radio specialists of McGraw-Hill publications***

***—to give most complete, dependable coverage of facts needed by all whose fields are grounded on radio fundamentals***

***—available at a special price and terms.***

These books cover circuit phenomena, tube theory, networks, measurements, and other subjects—give specialized treatment of all fields of practical design and application. They are books of recognized position in the literature—books you will refer to and be referred to often. If you are a researcher or experimenter—if your interest in radio is deep-set and based on a real desire to go further in this field—you want these books for the help they give in hundreds of problems throughout the whole field of radio engineering.

**5 volumes, 2981 pages, 2000 illustrations**

1. Everitt's COMMUNICATION ENGINEERING
2. Terman's RADIO ENGINEERING
3. Chaffee's THEORY OF THERMIONIC VACUUM TUBES
4. Hund's HIGH-FREQUENCY MEASUREMENTS
5. Henney's RADIO ENGINEERING HANDBOOK

### Special Low Price and Easy Terms

Bought singly, the five volumes comprising this library would cost you $26.00. Under this offer you save $2.50 and, in addition, have the privilege of paying in easy installments beginning with $2.50, 10 days after receipt of the books, and $3.00 monthly thereafter. Already these books are recognized as standard works that you are bound to require sooner or later. Take advantage of these convenient terms to add them to your library now.

SEND THIS ON-APPROVAL COUPON

McGraw-Hill Book Co., Inc.
330 W. 42nd St., New York, N. Y.

Send me Radio Engineering Library 5 vols., for 10 days' examination on approval. In 10 days I will send $2.50, plus few cents postage, and $3.00 monthly till $23.50 is paid, or return books postpaid. (We pay postage on orders accompanied by remittance of first installment.)

Name ..............................
Address ..............................
City and State ..............................
Position ..............................
Company .............................. RB-35

# Neutralizing a Transmitter With a NEON GLOW BULB

THE usual procedure for neutralizing a low power transmitter is to hold a flashlight bulb or small neon bulb near the tank coil and make adjustments until the neutralization is correct. This is fine so far as it goes but a much more convenient unit built for this purpose is shown in the accompanying illustration.

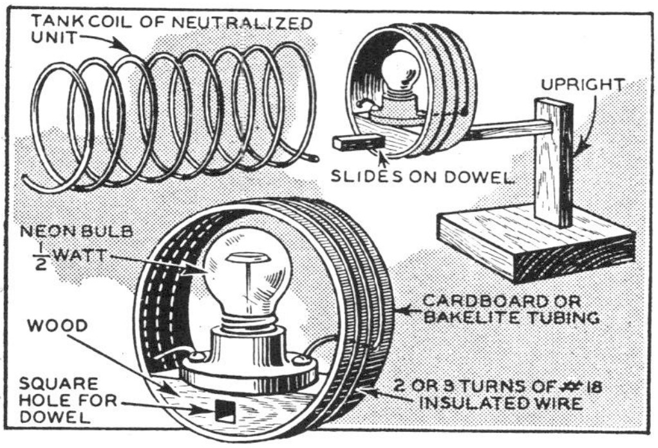


When neutralizing transmitter, slide the coil along the wooden rod till the correct point of adjustment is reached.

Cut out a section of bakelite or heavy cardboard tubing about three inches in diameter. Make the section about 1½ inches wide. Fit a small segment of wood to the inside as indicated and to its surface fasten an electric bulb socket into which can be screwed a small size neon bulb of about ½ watt.

Then cut a squared hole through the wood below the socket as shown. Run three turns of No. 18 insulated wire around the tubing, cementing it down with Duco cement and spacing the turns about two diameters. Drill holes for the leads in the bakelite and attach the ends of the turns to the socket terminals.

The standard is fitted with a square dowel projecting horizontally to center with the transmitter tank coil. The coil can then be slid on the dowel and adjusted to any desired distance from the end of the tank coil.

## Locating Bad Phone in Headset

WHEN a pair of headphones goes dead the first thing to hunt for is a burned out magnet coil. A simple way is shown in the sketch.

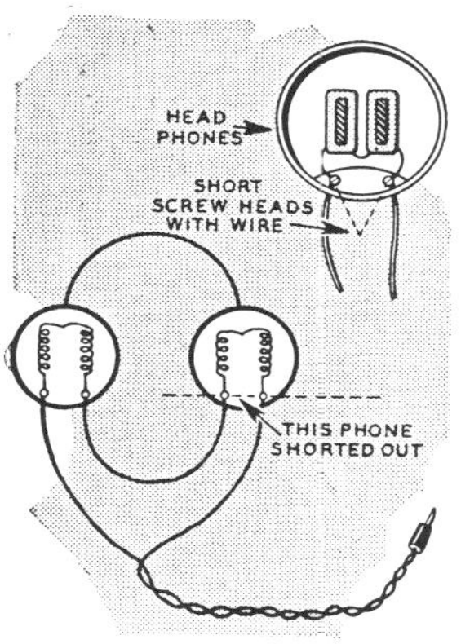


Short out bad phone with wire across terminals inside case.

Invariably it is only one phone that has gone bad. Being connected in series the circuit is, of course, broken. So the best method is by elimination. This is accomplished by shorting out first one phone and then the other.

Remove the earpiece and diaphragm from one phone. Hold a piece of metal across the two connector screws inside the case. Have the receiver running and place the other phone to the ear. If, when the remaining phone is shorted, sound comes into the other one you will know you have found the bad phone.

## Splitting Up a Headphone Set

WHEN there is a shortage of headphones, one good set may be split up, using a single headpiece for each listener. This will work all right if the unused ear can be "damped" by a suitable pad.

Wooden disks well padded with soft rubber or felt will take the place of the missing headsets and keep out external noises.

## Rubber Tubing Prevents Shorts

TO AVOID fireworks demonstrations and danger of burning up apparatus or bare flesh, radio experimenters can slip pieces of rubber tubing over unused high voltage terminals of experimental apparatus. Cut off sections of rubber tubing about twice as long as the terminals, and push in place. Where wire leads are in place the tubing may be slit lengthwise.

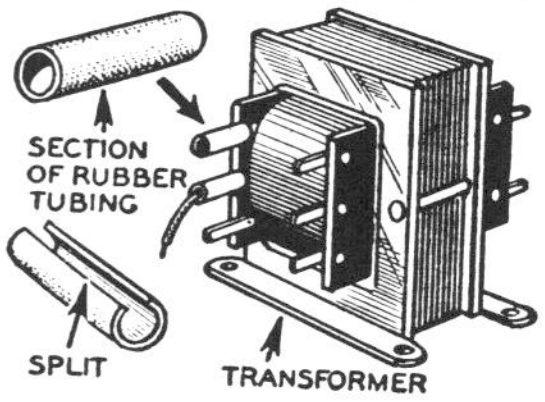


## Celluloid Chips Show Coil Sizes

TUBE base plug-in coils are apt to get mixed up with other short wave apparatus at times, and turns must be counted to determine the wave length for which each coil was wound. Ordinary poker chips glued or cemented to the top of the coil form, with the frequency or wave length of the coil clearly lettered, solve this trouble.

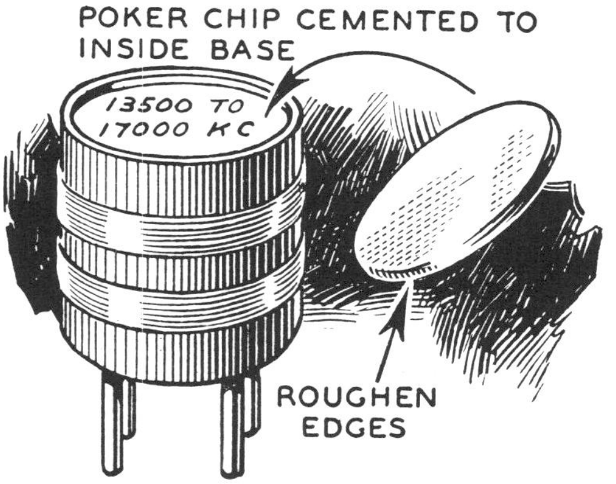

## Crystal Set Built Into Headphone

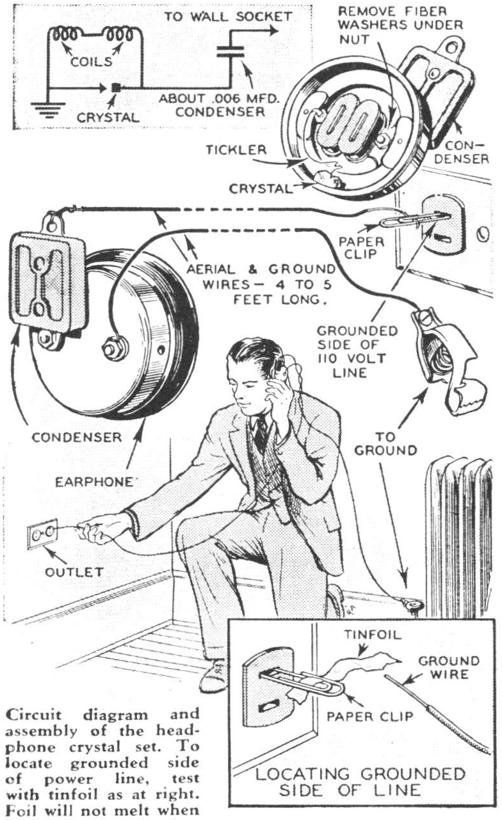


**Circuit diagram and assembly of the headphone crystal set. To locate grounded side of power line, test with tinfoil as at right. Foil will not melt when clip is at ground point.**

SNAP one lead of this headphone radio receiver to a ground pipe, push the other into a convenient electric socket, and enjoy the programs of your local broadcasting station.

This complete crystal radio receiver is easily built into an ordinary headphone; though only one station can be received with a given set of parts, this station may be changed by varying the value of the external fixed condenser.

Assemble the set as shown in the sketches. Use liquid solder in mounting the crystal, or a standard crystal holder, since any heat will damage sensivity. Remove the insulating washers from the indicated phone terminal to ground one end of the coil to the frame.

The grounded side of your house power line is used as aerial, and a radiator or water pipe as ground. Better results will of course be obtained with a regular outdoor antenna.

Cover and diaphragm must of course be removed when first adjusting the crystal. Once a sensitive spot is found, a few taps will often reset the catwhisker.

BUILD *Your own* RADIO *and Radio Devices*

You, too, can successfully build the many radio sets and radio devices described each month in practically every leading radio and technical publication.

Here at the world's Largest Radio Mail Order House you can buy approved, quality parts, kits and accessories at lowest WHOLESALE prices. Get our great 136 page Spring and Summer Catalog listing over 50,000 items. FREE Parts Lists on scores of radio devices are available. The radio kits briefly described below are typical of the many which we can supply.

Be sure to write us regarding your radio requirements before you start to build.

LOWEST WHOLESALE PRICES

**LITTLE GIANT 5 Tube AC-DC**

A powerful, compact superhet that is easy to build and highly efficient. Complete kit of approved parts with speaker, less tubes .......... $10.50

**BAND SPREAD PORTABLE**

A wonderful little S. W. set which is truly portable. Easy to build and very efficient. Entirely self contained. Kit of approved parts, less tubes, coils, batteries ......... $7.95

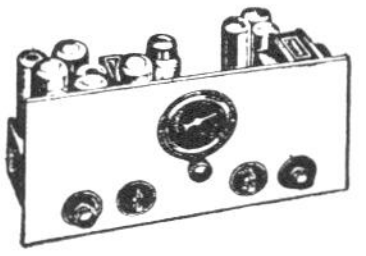

**ALL STAR Sr.**

A famous set, 6 tubes, ALL-wave Superhet, Tunes 10-500 M. Foreign reception guaranteed. Every latest feature. Kit of essential parts.... $32.14

**ALL STAR Jr.**

The latest 5 tube ALL-wave Superhet. Features ease of construction plus highest efficiency. Kit of essential parts, less tubes and speaker ........ $26.76

*When ordering from this advertisement or making inquiries regarding any kit, please address* Dept. IMH.

WHOLESALE RADIO SERVICE Co. INC

NEW YORK, N.Y. - 100 Sixth Avenue
ATLANTA, GA. 430 W. Peachtree St.

**Have you seen the big 1935 HOW TO BUILD IT? At your newsstand--50c.**

# SAVE MONEY
## WITH YOUR WORKSHOP TOOLS

DO YOUR OWN HOME and AUTO Repair Work.

◈◈

BUILD Furniture, Radios, a Camp Trailer, Boats, Novelties, and dozens of other big projects.

# LEARN HOW

*in the* **1935**

# HANDY MAN'S HOME MANUAL

50c AT ALL NEWSSTANDS    IF YOUR DEALER IS SOLD OUT SEND 50c TO

Modern Mechanix Publishing Co., 529 South 7th St., Minneapolis, Minn.

# Easily Made Rigid Coils for the Low Power Transmitter

FOR the amateur who builds his own transmitter coils, this method will aid in building very rigid windings, although No. 8 or 10 wire is used instead of the copper tubing as often specified. Rigidity is necessary to insure steady frequency, so here is a simple way to go about making such a coil.

Wind the coil in the usual manner and space the turns as near as possible to the distance needed. Now, depending on the diameter of the coil, cut from cigar box wood two or three slender sticks for spacers.

With calipers mark off the centers of the spaces for as many turns as the coil contains. Clamp the pair or the three sticks in a vise and cut the notches half the width of the wood.

### Thread Holds Windings in Place

Place a stick at a time inside the coil and temporarily tie it up to the windings. Then take a length of heavy black thread and tie one end to an end of the stick.

Take a turn over the end turn of the coil, down under the stick and then two more turns over the wire in the opposite direction and then under the stick up over the second turn. Make a second loop over this turn and then under the stick and up over the third turn, thus sewing the coil turns to the stick with a pair of thread loops over each turn.

Tie the thread to the opposite end of the stick and then with a blunt pointed needle pick up the turns over each coil turn, pulling taut, and repeat until the end of the coil has been reached. Finally take up the slack and retie.

This should effectively tie all coil turns solidly down into the stick notches. Repeat with the remaining sticks. When this is finished, spread a solid coating of Duco Cement along the entire length of the stick on the outside of the coil as shown and allow to dry several hours. When dry the coil will be as rigid as though built of one piece and can be handled without fear of bending it out of shape.

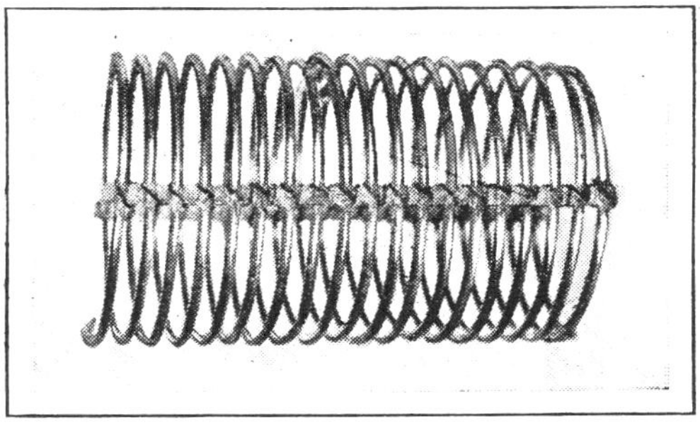

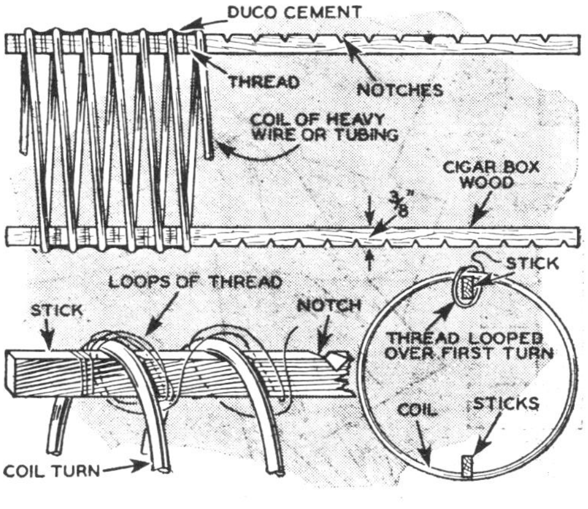


Top photo shows completed coil. Below is shown the manner in which coils are wound on sticks.

# Arranging the Automatic Grid Bias When Converting DC Receiver to AC

MANY of the older style battery receivers using C batteries for grid bias are now being converted to AC operation and a resistance is used for obtaining this bias without the use of batteries.

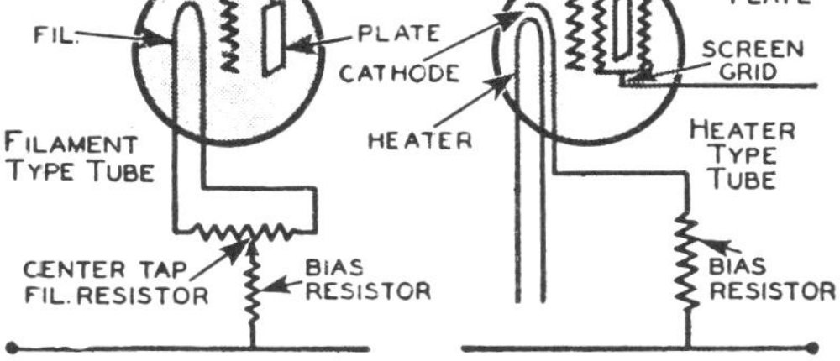


Bias resistor is wired in at points shown in above drawing when direct current set is changed over to alternating.

The resistance is always inserted between the filament return and the B minus, as illustrated herewith. In the case of a purely filament heated tube this is accomplished by shunting the filament circuit with a 50 to 100 ohm resistor and center tapping that. Then insert the resistance in the center tap lead to B minus.

In the heater type tube the cathode leads direct to the B minus and the resistance should be inserted in that lead. Estimating resistance is simply a matter of experimentation and of knowing the characteristics of the tube used.

These characteristics are to be found on the pamphlet accompanying the tube. With these changes you can increase the efficiency of your receiver considerably.

# FREE

MODERN MECHANIX PUBLISHING CO.
529 S. 7th St., Minneapolis, Minn.

Please rush my free copy of the big 1935 MM Blueprint catalog. I am enclosing a 3-cent stamp to cover postage.

Name ..............................................

Address ..............................................

City ..............................................

State ..............................................


## Winding Filament Transformers

| WATTS | TURNS PER VOLT |
|---|---|
| 5 | 20 |
| 10 | 15 |
| 15 | 12 |
| 20 | 10 |
| 25 | 9 |

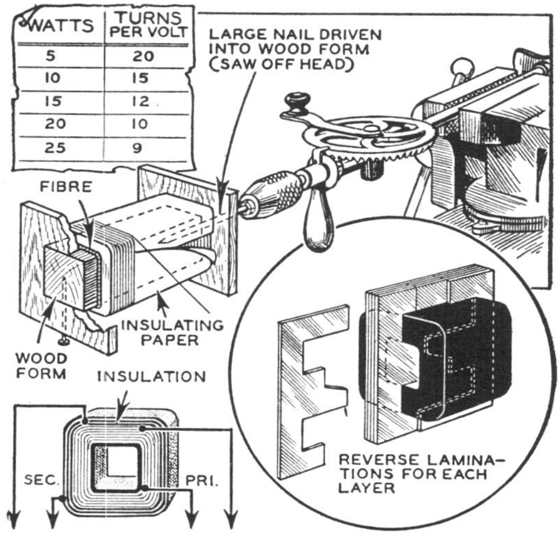


Add 10% to filament wattage for losses to get primary watts. Multiply turns per volt given above for needed wattage by voltages to get turns per winding. Compute currents, and select correct enameled wires from wire table.

THE cores of old audio transformers are often large enough for filament transformers. Use the fibre core of the old winding as a coil form, winding the primary on first. Run the winding across, but do not mind if the turns pile up on each other. Place a thin sheet of wax paper between each layer, and heavier insulation between primary and secondary. Measure the output voltage with a voltmeter, adding or subtracting turns to get the correct voltage when filament is connected.

Make a wood coil form and mount it in a breast drill held in a vise. A large nail is driven into the center of the form, as shown.

## Test Condensers With Light Bulb

A TEST light is always handy on the radio workbench. With it you can make many continuity tests, and even check condensers. Connect a pair of test leads with clips to a bayonet plug, placing lamp socket in series with one of the leads. To test condensers plug into a 110 volt source, and connect clips to the condenser. If the bulb lights up the condenser is shorted. If it flashes for an instant, glows faintly, or is dark, the condenser is good.